Robert A Lewis

Fisheries and Energy Production: A Symposium

Fisheries and Energy Production: A Symposium

Edited by

Saul B. Saila
University of Rhode Island

Lexington Books
D.C. Heath and Company
Lexington, Massachusetts
Toronto London

Published simultaneously in Canada

Printed in the United States of America

International Standard Book Number: 0-669-98467-1

Library of Congress Catalog Card Number: 74-32513

CONTENTS

FOREWORD

Adequate electrical energy supplies and power plant sites are important for the future of our country. Adequate electrical energy supplies raise difficult environmental questions--some of which fishery scientists can help answer.

This symposium "Fisheries and Energy Production" has the following objectives:

1) to make more available to interested persons some of the material related to this subject;

2) to demonstrate some of the progress which has been made in assessing and reducing some of the fisheries problems related to current methods of energy production; and

3) to demonstrate some of the methods utilized to assess the conditions of the aquatic environment as they relate to fisheries.

The symposium is limited in scope and does not encompass all the fisheries problems related to energy production. However, an effort has been made to include elements of the entire aquatic food web and to include topics of current interest. The size of the symposium volume and the strict publishing schedule has restricted the number of contributors.

The financial support for the editing and publication of this symposium came from two sources: the United States Atomic Energy Commission and the United States Environmental Protection Agency. The personal encouragement of D. Heyward Hamilton, Jr., Environmental Programs, Division of Biomedical and Encironmental Research, U.S.A.E.C. and Andrew McErlean, Office of Research and Development, U.S.E.P.A. are especially appreciated.

Deborah Van Dam has helped significantly by reviewing, criticizing and generally improving the manuscripts as they have been received.

Mrs. Janice B. Steele has retyped and proof-read the edited manuscripts.

The president of the Northeast Division of the American Fisheries Society, Dr. Alfred Eipper, has been an enthusiastic supporter of the symposium since its beginning. The American

Fisheries Society, through Dr. Raymond Johnson, has also considered this a worthwhile project.

The authors of the various chapters are also to be thanked for their cooperation in complying with a very tight time schedule.

The editor assumes responsibility for the errors of substance or grammar which might remain.

There is a great need for improved communication, and members of the scientific community, including fishery scientists, can greatly assist in the process. Fisheries and energy production is a complex subject. This symposium represents a small step in alleviating the communications problem by drawing upon the experience of experts and making it available to anyone.

Fisheries and Energy Production: A Symposium

SIMULATING THE IMPACT OF THE ENTRAINMENT OF WINTER FLOUNDER LARVAE

K. W. Hess, M. P. Sissenwine and S. B. Saila
Marine Experiment Station
University of Rhode Island, Kingston, Rhode Island 02881

ABSTRACT

The transport of winter flounder larvae around the Millstone Point, Conn. Area by the action of tidal currents and diffusion was simulated by computer to predict the numbers which could be entrained during the operation of a local nuclear power station. A tidal hydrodynamic model with variable depth was employed to simulate currents and water levels. These techniques provided input to a transport model which simulated the concentration of larvae. A larval source in a tributary river was simulated for twenty tidal cycles, with and without entrainment. The results indicated that the reduction in winter flounder larvae near Millstone Point at the end of the pelagic stage (period during which larvae are likely to be entrained) was less than 1% when it was assumed that larvae have little chance of returning once lost from Millstone bight.

In order to assess the effect of a 1% reduction in recruitment of winter flounder larvae to the benthic phase of their life cycle, the local population was simulated by a model in which year-classes and the total egg production were represented by compartments. Each year-class grew, produced eggs, suffered natural and fishing mortality according to information derived from the literature. The effect of power plant entrainment was incorporated by reducing the number of recruits to year-class I that would normally result from a specific level of egg production. For a 1% reduction in recruitment due to power plant entrainment, a potential 6% decrease in total population size following 35 years of operation was indicated.

INTRODUCTION

The Millstone Point nuclear power station is located on the north shore of Long Island Sound and east of the Niantic Bay and River (Fig. 1). Three nuclear power plants equipped with once-through cooling systems requiring about 115 m^3/second of cooling water are anticipated by 1981. The cooling water will contain small planktonic organisms commonly found in the marine environment. These organisms are stressed by rapidly changing hydrostatic and thermal conditions. This report considers the effect

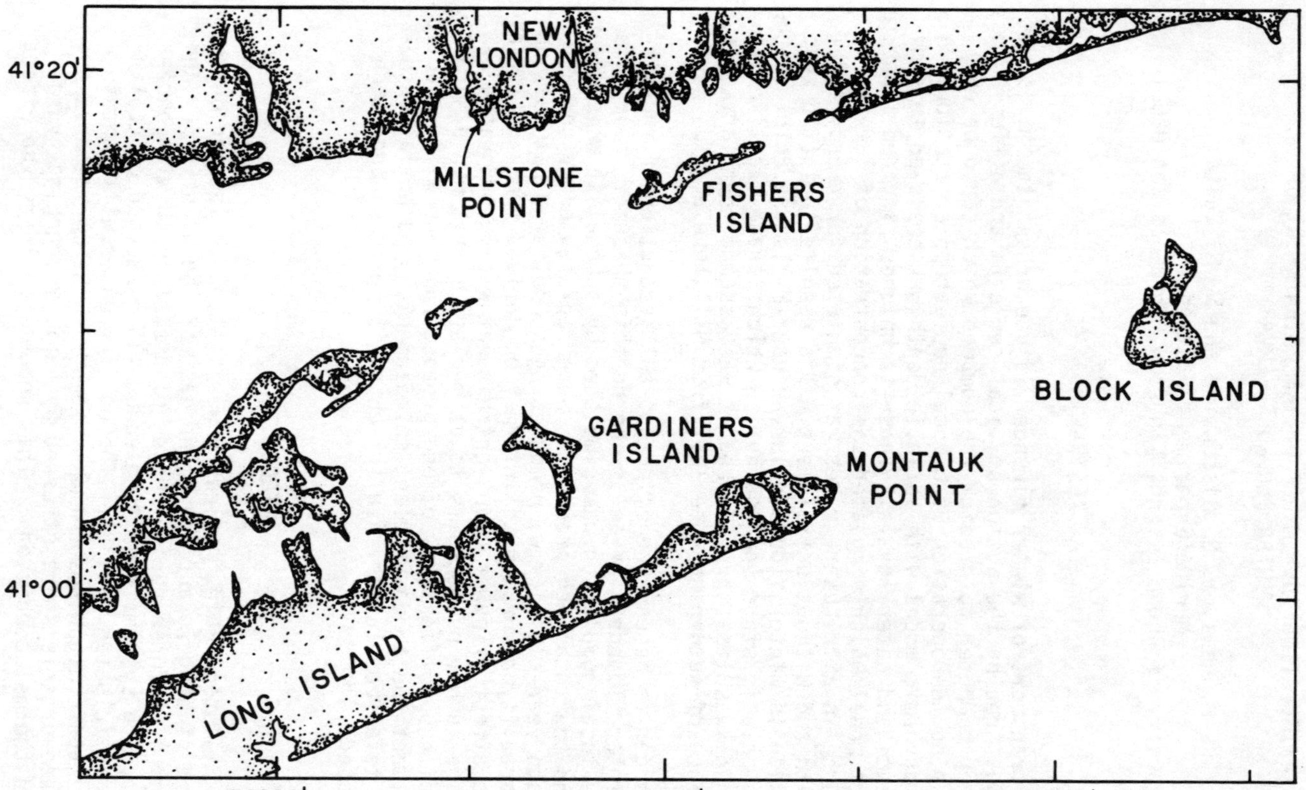

Figure 1. Eastern Long Island Sound.

of entrainment at the population level assuming organisms passing through the plant suffer 100% mortality.

Larvae of the winter flounder, Pseudopleuronectes americanus, are found in the water column near Millstone Point during the spring of each year and are therefore subject to entrainment. The entrainment of these pelagic larvae decreases the number of organisms recruited to the benthic winter flounder population.

Winter flounder usually breed in shallow semi-isolated inlets and estuaries such as the Niantic River. The preliminary results of a trawling survey being conducted in the region indicate that mature winter flounder are abundant in the Niantic River during their late winter and early spring spawning season. While the adult fish leave the river and move into deeper water as summer approaches, they probably do not migrate far from the breeding area (Lobell, 1939; Perlmutter, 1939). Early tag returns of fish captured and released in the Niantic River support this contention. Saila (1961a) demonstrated the ability of adults to return to a specific breeding site in large numbers. Based on the available information, it appears reasonable to assume that the Niantic River winter flounder population is reproductively isolated, although some exchange of larvae and adult fish with nearby populations occurs.

Isolated populations are particularly sensitive to localized environmental stresses since, for example, losses by entrainment of the Millstone Point Nuclear Power Station are not easily offset by input from other populations. Therefore, it is important to assess the impact of the entrainment of winter flounder larvae on the Niantic River population.

The effect of entrainment will be called "plant mortality." It acts in addition to natural mortality, so that fewer winter flounder larvae survive to their benthic stage with the power plant in operation. In this study, the effects of natural and plant mortality are separated to simplify calculations. By considering the sources of mortality separately, the possibility of a decrease in natural mortality to compensate for entrainment losses is ignored. Thus the approach tends to overestimate the impact of entrainment.

First, a hatch of organisms in the Niantic River and other possible breeding areas is simulated along with the movement of the larvae under the action of tidal-currents around the river, the waters near Millstone Point and into Long Island Sound. A comparison of the final number in the waters near Millstone Point at the end of the larval period, with and without the plant in operation, is used to estimate the entrainment mortality rate. Then this result is employed in a compartmental model of the winter flounder population in order to simulate the impact of entrainment after many years.

The Local Dispersal of Winter Flounder Larvae

The transport of larvae around the Niantic Bay area is accomplished primarily by advection due to currents and secondarily by diffusion based on concentration gradients. Movement by the larvae themselves is ignored because of their poor swimming ability. The net displacement per unit time of winter flounder larvae in laboratory chambers was measured. Using these results, the self-diffusion rate of the larvae was calculated, based on a random walk model, and found to be several orders of magnitude lower than the turbulent diffusion rate. The dominant component of the prevailing circulation is that due to the tides in Long Island Sound. Wind-induced currents also produce short-term perturbations which are of only secondary importance in the total circulation. Gravitational or density-driven currents may exist only in the upper reaches of the Niantic River where noticeable salinity gradients result from brook discharge.

The currents in Niantic Bay sweep the larvae around the bight and past the cooling-water intake structures of the nuclear power station, which themselves augment the local circulation. The fate of larvae hatching in the Niantic River and other areas of the bight can be studied by using the currents as input to solve a convective-diffusive equation of mass conservation. The time-varying spatial concentration of organisms can be predicted, and the effect of the power plant assessed.

The tidal circulation in the bight is modelled numerically using a horizontal grid network consisting of 287 internal square grids 305 meters on a side (Fig. 2). The time-dependent equations of flow and mass conservation in a fluid, assuming constant density and a hydrostatic vertical pressure relationship, are employed. These equations are integrated over depth from the bottom to the water surface to give, for momentum,

$$\frac{\partial U}{\partial t} + U\frac{\partial U}{\partial x} + V\frac{\partial V}{\partial y} = - g\frac{\partial \eta}{\partial x} + fV + \frac{1}{\rho(h+\eta)}(\tau_{sx} - \tau_{bx}) \qquad (1)$$

$$\frac{\partial V}{\partial t} + U\frac{\partial V}{\partial x} + V\frac{\partial V}{\partial y} = - g\frac{\partial \eta}{\partial y} - fU + \frac{1}{\rho(h+\eta)}(\tau_{sy} - \tau_{by}) \qquad (2)$$

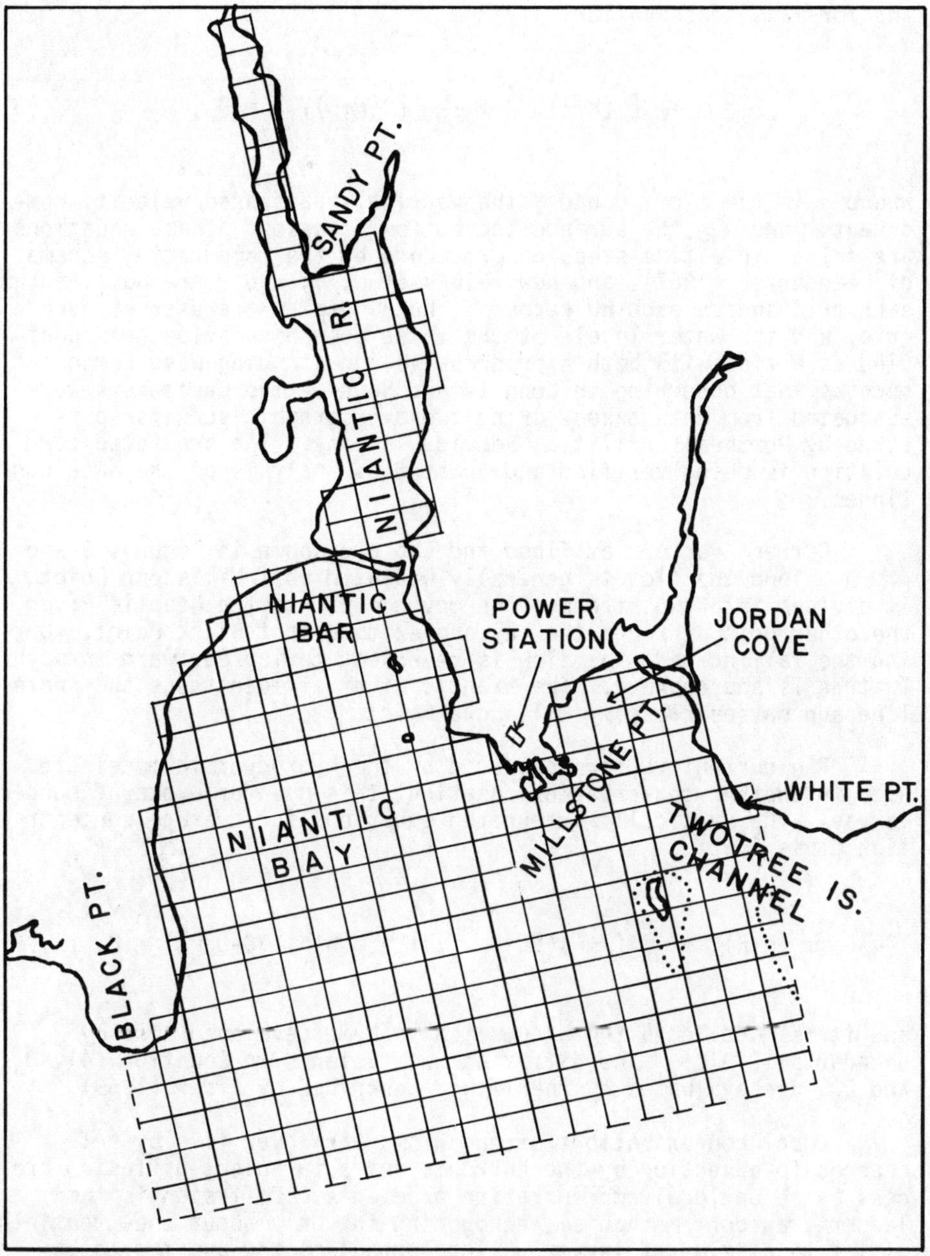

Figure 2. The grid network of the circulation and dispersion models of the bight.

and for mass conservation

$$\frac{\partial \eta}{\partial t} + \frac{\partial}{\partial x} [(h+\eta)U] + \frac{\partial}{\partial y} [(h+\eta)V] = 0 , \qquad (3)$$

where η is the tide, U and V the vertically-averaged velocity components, and τ_{ab} the surface and bottom stresses. These equations are solved in a time-stepping procedure by the computation scheme of Leendertse (1967), and new values of U, V, and η are obtained at each grid-square each 60 seconds. Local depths are used at each grid, and the water levels at the three open boundaries are specified as a tide with both a progressive and standing wave component such as that occurring in Long Island Sound. The parameters were evaluated from data taken during two hydrographic surveys sponsored by Northeast Utilities Service Company. The predicted circulation is being verified and updated as analysis of the data continues.

Current vectors at flood and ebb are shown in Figures 3 and 4. At flood the flow is generally westward past Millstone Point. It divides into two streams, one moving up into the Niantic River, the other going through the bay and exiting past Black Point. During the falling tide the flow is reversed, moving eastward through Twotree Island channel. The Niantic River effluent hugs the shoreline and passes close by Millstone Point.

The current vectors produced by the hydrodynamic model are used in another conservation equation, this one for winter flounder larvae. The vertically-integrated equation, for average concentration C, is

$$\frac{\partial}{\partial t} (h+\eta)C + \frac{\partial}{\partial x} (H+\eta)(UC-D_x\frac{\partial C}{\partial x}) + \frac{\partial}{\partial y} (h+\eta)(VC-D_y\frac{\partial C}{\partial y}) = 0 \qquad (4)$$

and it is modelled using a numerical scheme developed also by Leendertse (1970). The diffusion coefficients in Equation (4), D_x and D_y, are evaluated by the method described by Elder (1959).

Local concentrations of organisms vary over time by net changes in advection by the currents and a turbulent diffusion process based on local concentration gradients. Diffusion is the lesser transport mechanism, accounting for only about one-twentieth of the total flux of larvae. Since organisms are assumed to be non-swimming, they move with the currents. The present model

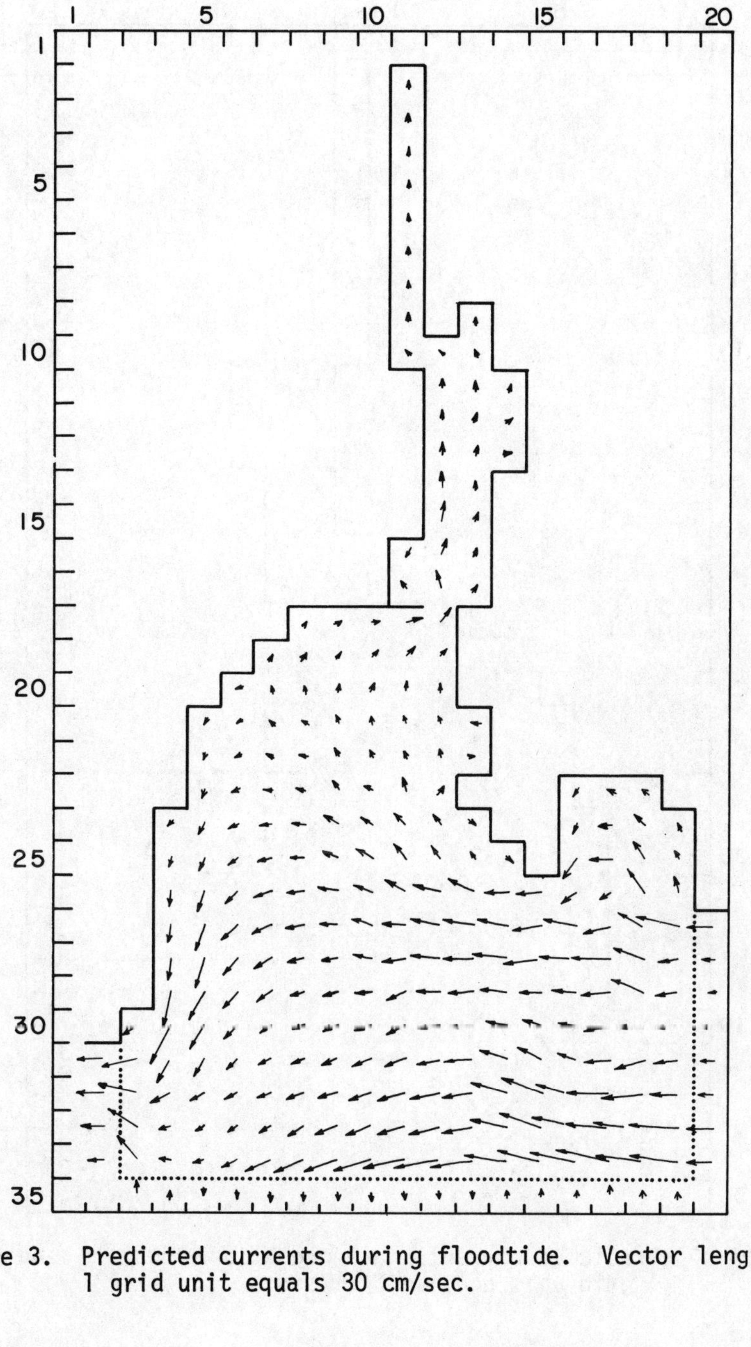

Figure 3. Predicted currents during floodtide. Vector length of 1 grid unit equals 30 cm/sec.

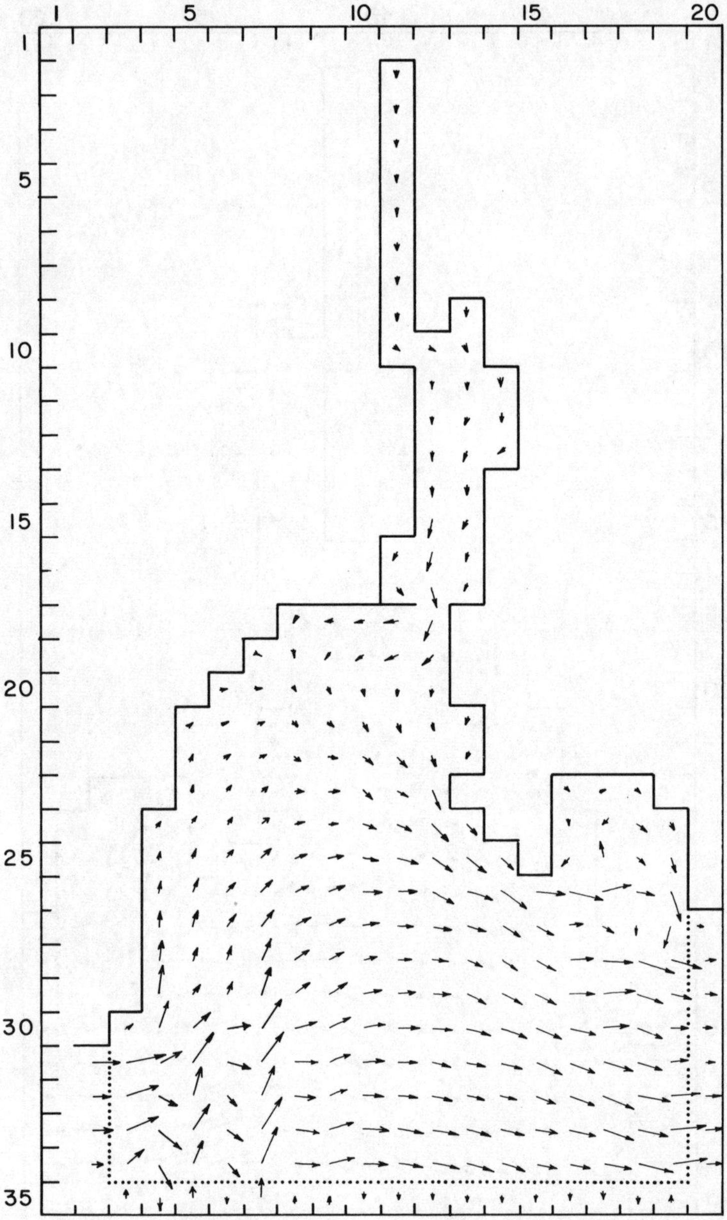

Figure 4. Predicted currents during ebbtide. Vector length of 1 grid unit equals 30 cm/sec.

assumes a uniform vertical distribution of larvae in the column. Non-uniform configurations and vertical migration can be introduced with minor modifications.

The effects of entrainment can be modelled by adding a sink at the grid boundary representing a river-like larval loss at the Millstone Point location. Larvae crossing the outer boundaries of the modelled region are assumed to be lost from the system. This reflects the inhospitable conditions encountered there by the organisms. Likewise, no larvae are assumed to enter the region from the outside, so that the local larval population is the only one considered.

A series of simulations was carried out to model the fate of flounder larvae hatched in several areas of the greater Millstone bight. The results indicated that the effects of entrainment on these organisms is small.

The first region of hatch extends from the mouth of the Niantic River north to the Sandy Point area. This region includes broad sandy bottoms which are favored by winter flounder for spawning. The lower river does not have this ideal bottom environment, so it should reflect rapid larval loss and an earlier indication of larval dispersal in the Niantic Bay proper.

Another region likely to host hatching organisms is the upper arm of the Niantic River north of Sandy Point. In fact, many individuals from the first hatch are transported to this location (according to the results of the computer simulation), and remain there for a long period of time. This region appears to be the most favorable for the retention of larvae in the estuary. The probable fate of these organisms when entering Niantic Bay is virtually the same as for organisms hatched in other parts of the river, although it may occur several tidal cycles later.

Two bay hatch areas were also modelled. These were near the Niantic bar and in outer Jordan Cove. The larvae from these regions were rapidly transported out of the bight.

During the first several tidal cycles of the simulated river hatch there was rapid loss of larvae from the river. This is attributable to the facts that the lower river is not well suited to larval retention, and that a large gradient between river and bay is present, promoting greater diffusion. Over one simulated tidal cycle, both with and without the effects of the power plant, about 72% of the larvae leaving the river at ebbtide returned to the river on the following floodtide (Fig. 5). The high probability of return may partially explain the strong larval retention

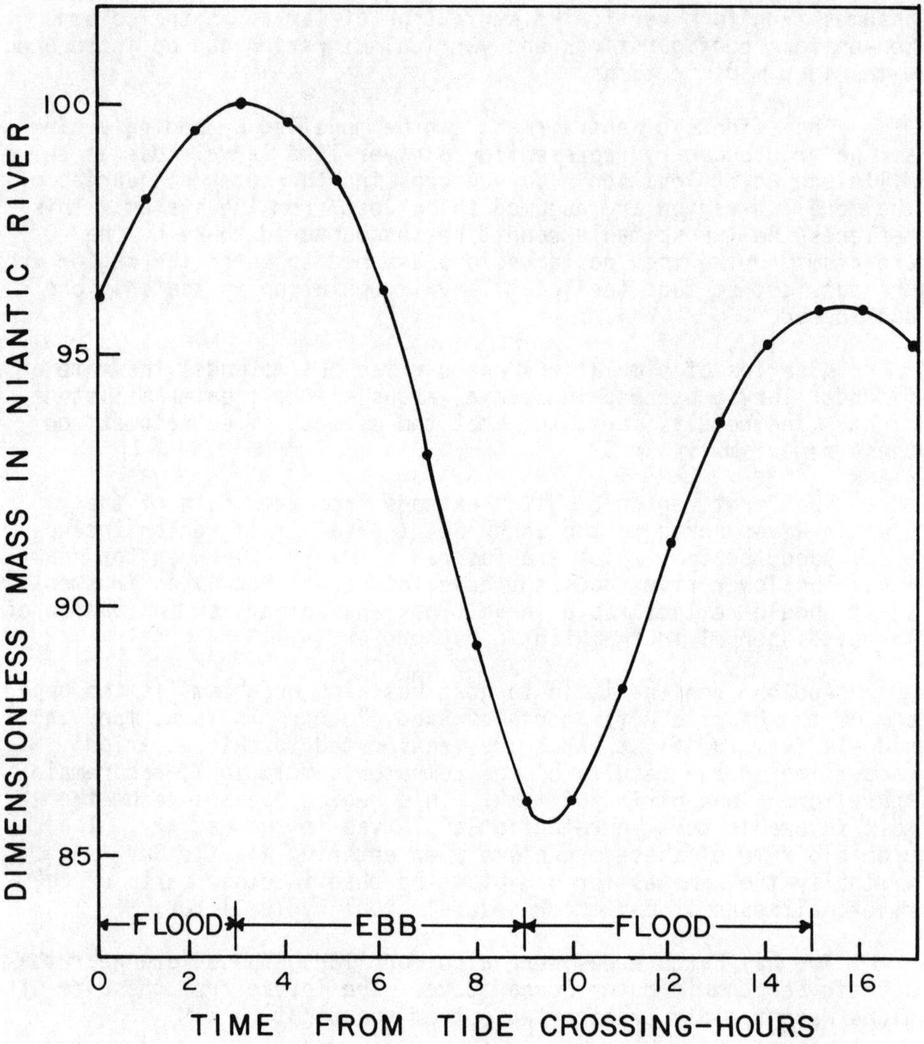

Figure 5. Simulated mass in the Niantic River, indicating that 72% of the mass leaving the river on ebbtide returns on floodtide.

characteristic of the estuary. Since there is little difference between the fraction returning for the two cases (with and without entrainment), it appears that potential entrainment mortality of organisms remaining in the river will be small.

The computer model of the Millstone bight has been used to assess the effects of entrainment on constituents in the waterway by employing an outflow at one grid square representing the plant intakes. Since the circulation is rather crudely modelled in this region another model was developed to predict the local currents on a finer scale. A grid network consisting of 326 squares, 61 meters on a side, was developed for this purpose. Tidal currents in this network were computed using water levels from the larger model as boundary input.

The three intake structures were modelled separately, and preliminary results indicate that the total computed masses lost through the intakes for each model (305 and 61 meters grid size) were approximately equal. Thus the details of circulation in the intake region are realistically incorporated in the larger grid size model.

Larval dispersal over twenty tidal cycles for a Niantic River hatch area south of Sandy Point was simulated. The distribution of concentration is shown in Figure 6. The patterns show that most of the organisms leaving the river progress toward Millstone Point and continue out of the modeled area via Twotree Island Channel, where they are assumed lost from the system. Initially some larvae enter the outer Jordan Cove area, but in the model they are rapidly flushed out. Thus there should be no build-up of organisms in this vicinity. After several tidal cycles the rate of larval loss diminishes, and the fraction of the total remaining larvae which stays in the river increases. This is indicative of the favorable retention characteristics of the river.

The fraction of the initial organisms remaining at the end of each of the first twenty tidal cycles is shown in Figure 7. The loss is rapid for the first 12 cycles, but appears to lessen beyond that. There is an initial build-up in the cove, which reaches a maximum after five cycles and then diminishes. Only 30% of the total organisms remain in the bight after twenty cycles.

A direct comparison of those remaining with and without the plant in operation is shown in Table 1. At the eleventh tidal cycle the percent difference in the two fractions is about 1.2%. At the twentieth cycle it has diminished to about 0.7%, and is decreasing. Therefore, for the 120 to 150 tidal cycles experienced

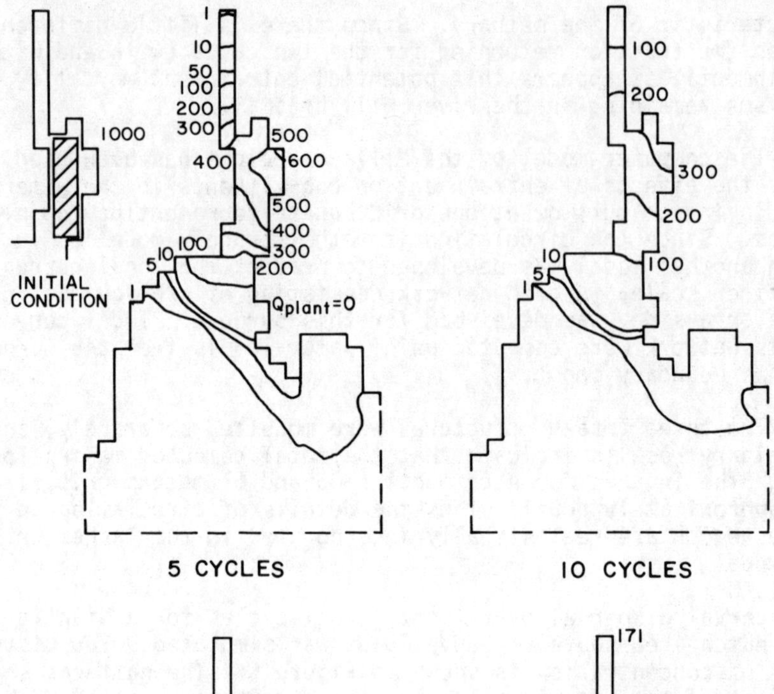

Figure 6. Simulated concentration distribution of larvae at 5, 10, 15 and 20 tidal cycles after hatching in Niantic River.

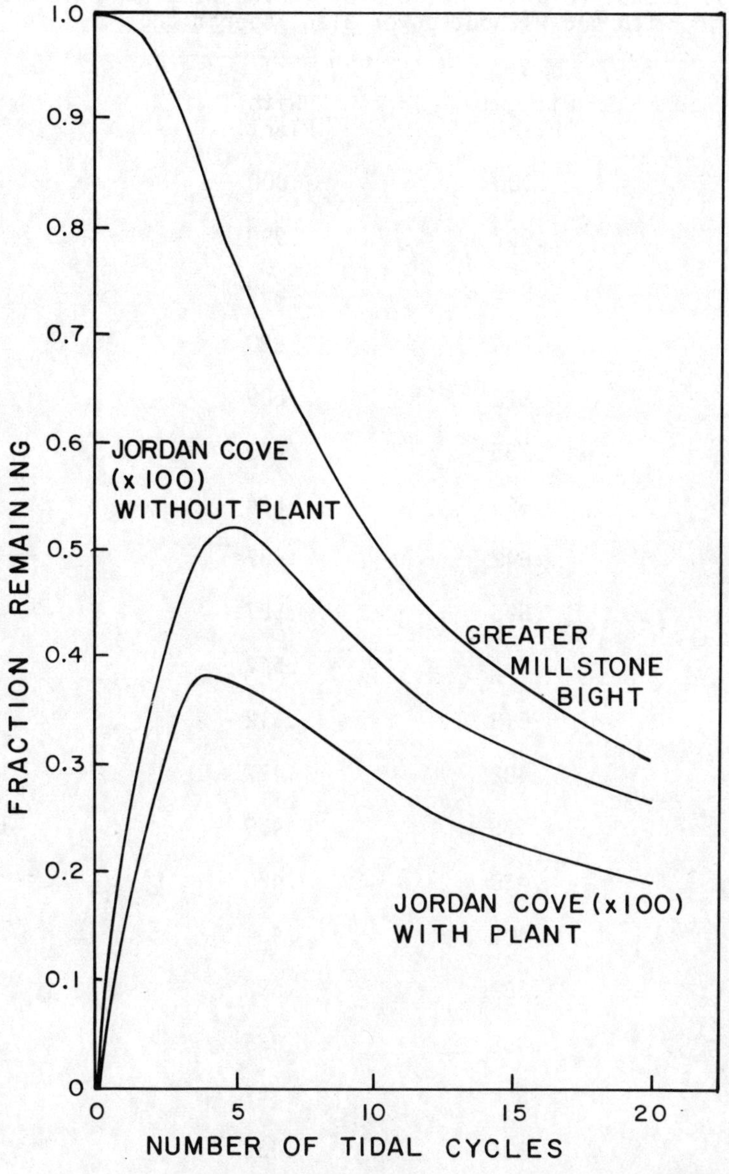

Figure 7. Fraction of organisms remaining in the greater Millstone bight and Jordan Cove. The fraction remaining in the bight with and without the plant differs only slightly, as indicated in Table 1.

Table 1. Total relative numbers of larvae remaining in the bight with and without power plant operation.

Tidal Cycle	Without Plant	With Plant	% Difference
0	1.000	1.000	-
1	.997	.993	-
2	.959	.948	-
3	.894	.880	-
4	.822	.809	-
5	.755	.743	-
6	.695	.684	-
7	.642	.632	-
8	.595	.587	-
9	.554	.547	-
10	.518	.512	-
11	.482	.477	1.180
12	.455	.450	1.076
13	.430	.426	.998
14	.408	.404	.957
15	.389	.383	.903
16	.368	.365	.848
17	.350	.347	.800
18	.334	.332	.748
19	.319	.317	.721
20	.305	.303	.688

by flounder larvae, projected plant mortality M_{pl} is probably less than 1%, based on the present results and model assumptions.

It has been shown that the Niantic River is ideally suited to be a nursery area because it impedes larval loss by limiting the flushing rate. The dispersal model indicates that the total mass of larvae in the river decreases by only 4% per tidal cycle (after 20 cycles) and this fraction slowly becomes smaller. Assuming a 4% loss rate for a total of 150 tidal cycles (or two and one-half months), the model indicates that about 250 of 100,000 larvae hatched in the river remain in the system long enough to reach their benthic stage. This number is adequate to account for the larval survival of the species. In contrast, the rest of the bay is flushed more rapidly.

A hatch area just south of the bar was simulated with the larval dispersal model (Fig. 8). After just four cycles, only 25% of the initial mass was left. This compares with 80% for a river hatch. It is interesting to note that a sizable fraction of the remaining larvae are in the Niantic River. The distribution of concentration is similar to that after a river hatch, indicating that the primary loss path is through Twotree Island Channel.

Another hatch area modelled was the outer Jordan Cove (Fig. 9). The rate of loss there was even more rapid. After a full tidal cycle only 19% of the initial mass was left. This indicates that the outer cove is well-flushed, and only the near-shore areas have a probability of retaining a larval population.

The results of present model studies indicate that larvae originating in the Niantic River remain in the bight for long periods of time, and that mortality inflicted by the power plant on organisms in the river is minimal. Larvae escaping from the river suffer some damage from the plans. In most cases they would be lost into Long Island Sound. Organisms hatching in other areas of the bight are more rapidly flushed from the bay. Present results show that power plant operation during the winter flounder larval stage would result in a reduction in flounder entering year-class one of less than one percent.

Winter Flounder Population Model

The structure of the winter flounder population model is represented by Figure 10. YC_j for j = 1,2,...,12 is the level (number or biomass) of the jth year-class. There is a continuous loss from each year-class by natural and fishing mortality. The loss rates are proportional to the level of the particular year-

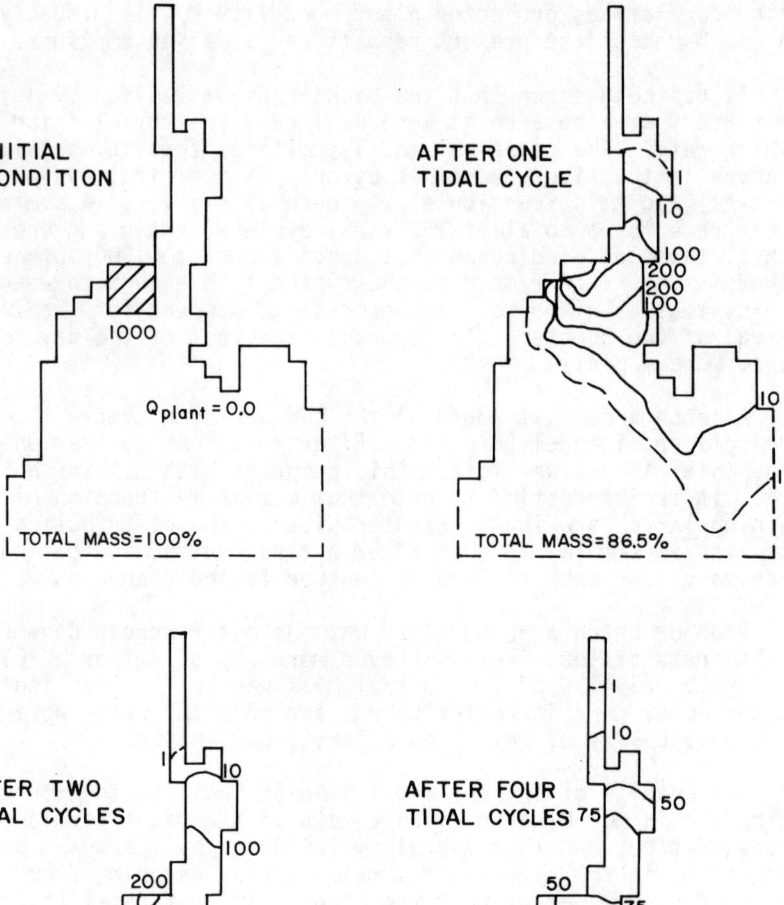

Figure 8. Simulated concentration distribution of larvae at 1, 2 and 4 tidal cycles after hatching near the Niantic Bar.

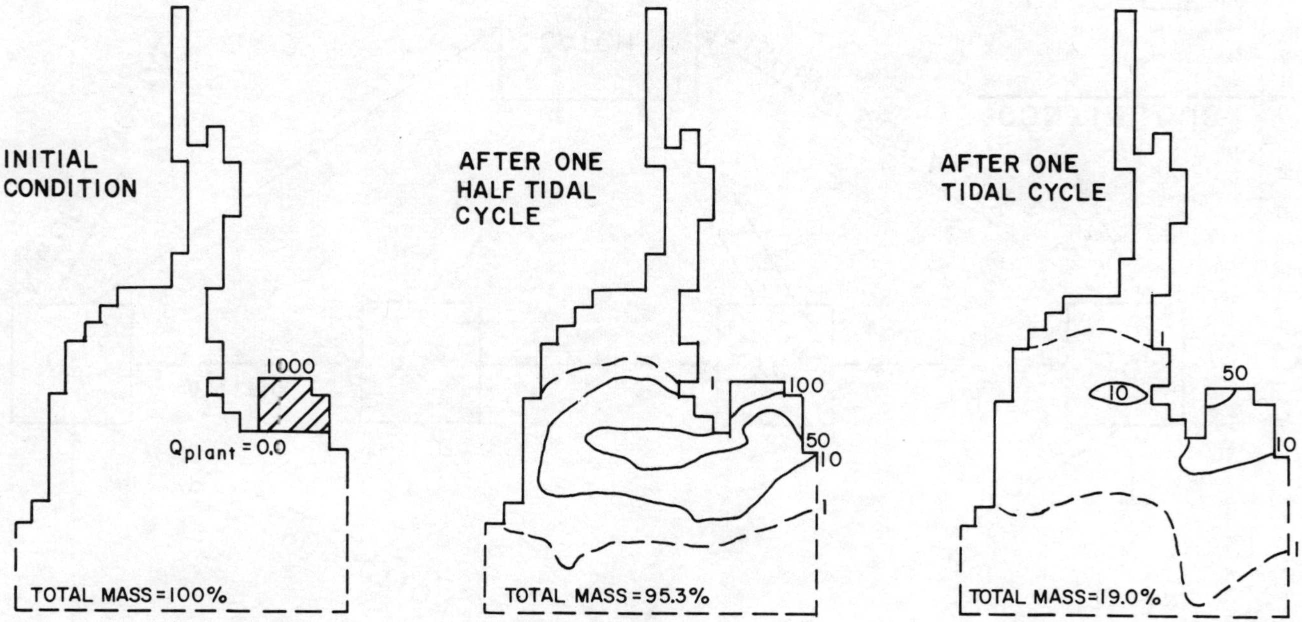

Figure 9. Simulated concentration distribution of larvae at 1 and 2 tidal cycles after hatching in Jordan Cove.

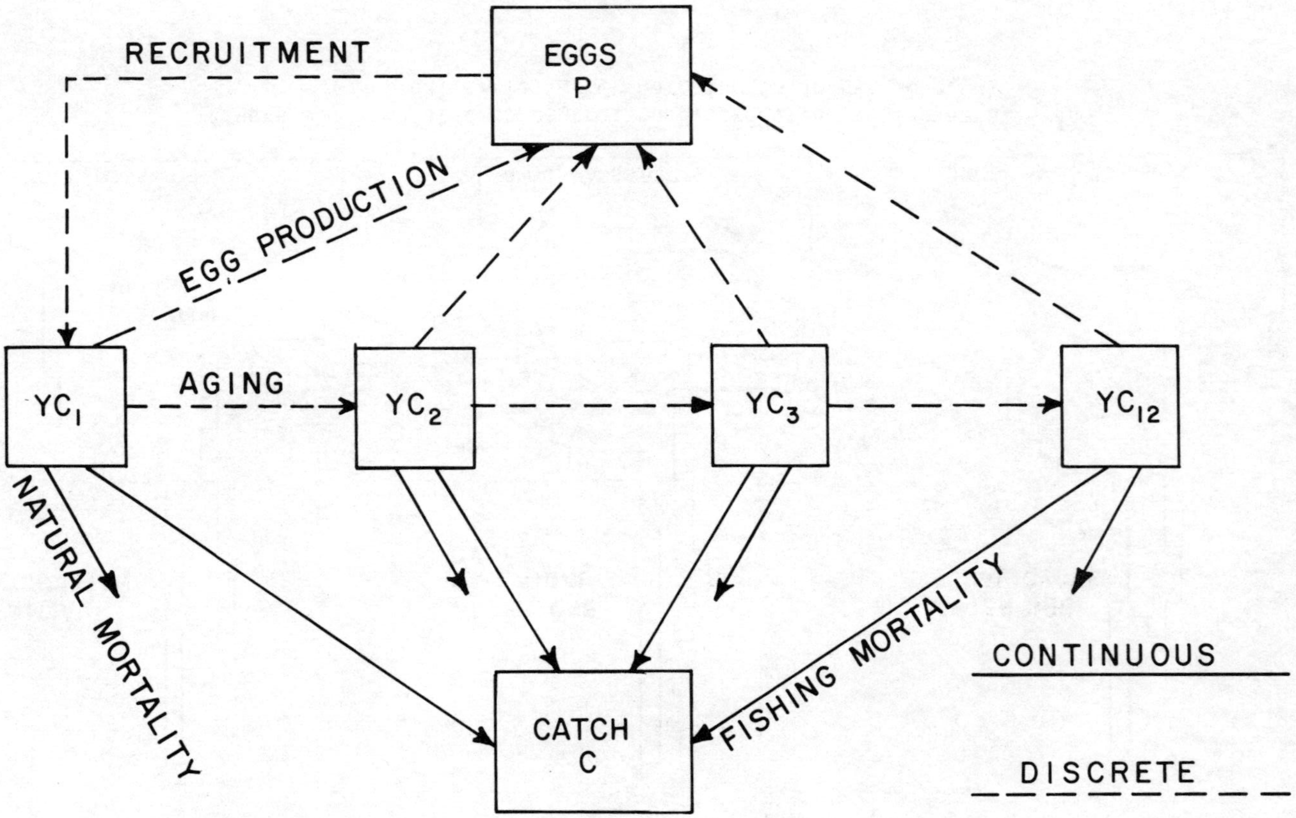

Figure 10. Winter flounder population compartment model.

class compartment with which they are associated. The loss of biomass due to fishing is added to the catch. At the beginning of each year, survivors of YC_j advance to YC_{j+1} and eggs are produced according to the mean fecundity of the year-classes. Recruitment to year-class 1 is a function of the number of eggs produced during the previous year multiplied by $1-M_{p1}$, where M_{p1} is the reduction in recruitment resulting from entrainment.

Figure 10 describes a closed system in which exchange with other populations is ignored. In this way the model is conservative, since some of the reduction in recruitment from entrainment might be offset by immigration as local populations decrease.

Application of the model required specification of the initial conditions and parameters regulating the system. Values of the fishing and natural mortality rate, the mean weight and fecundity of each year-class and the sex ratio were based on the available literature. Both the Ricker (1954) function and a density independent function were used to define the relationship between egg production and recruitment.

According to Saila (1961b):

$$\log FEC = 2.6712 + 1.1383 \log W \quad (5)$$

where FEC and W are the fecundity (number of eggs per female) and weight in grams of a winter flounder. The relationship between the length and weight of a winter flounder was reported by Lux (1969),

$$\log W = - 5.239 + 3.138 \log L \quad (6)$$

where L is in mm. The information in Table 2 was based on the mean length of winter flounder by age and sex (Berry, Saila and Horton, 1965), a sex ratio of 7 females to 3 males (Saila, 1961b), and Equations 5 and 6.

At present, there is no information available concerning the size of the resident winter flounder population in the Millstone bight. Saila (1961b) estimated the population density of yearling and older (age 2 and greater) winter flounder as 26.19 and 5.68 per $1000m^3$ for Charlestown and Green Hill Ponds, Rhode Island. Field studies to date have shown that winter flounder are concentrated in the Niantic River during the spawning season. Based on the area of the river (3,196,800 m^2), an initial population size of 83,724 yearling and 18,158 adult fish was assumed.

The mortality rate of adult winter flounder was assessed by Berry et al. (1965) and Poole (1969). Berry et al. estimated the

Table 2. Length, weight and fecundity of winter flounder by age and sex. L in mm and W in grams.

Age (Years)	Males L	Males W	Females L	Females W	Females FEC (x 1,000)	Average W
1	125	21.9	135	27.9	-	25.7
2	188	78.9	215	120.3	-	105.2
3	236	160.0	274	257.4	260	227.3
4	275	260.4	318	410.8	443	356.1
5	304	356.7	352	565.0	637	489.2
6	326	444.1	377	700.7	813	607.3
7	342	516.1	396	817.7	970	707.9
8	354	575.1	401	918.8	1107	793.7
9	363	622.3	422	998.2	1217	861.4
10	370	660.7	430	1058.8	1301	913.9
11	375	689.1	436	1105.9	1367	954.2
12	379	712.5	441	1146.2	1424	988.3

instantaneous total mortality rate as 0.92 (60% per year) based on the age-class structure of fish from Charlestown Pond and Narragansett Bay. Poole considered tag and recapture data from Great South Bay, New York and found an instantaneous total mortality rate of 1.38 (75% per year). Therefore instantaneous fishing and natural mortality rates of 0.45 and 0.66 were assumed for fish of year-class 2 and older; these values correspond to a total annual mortality rate of 67% per year and partitioning of mortality between sources in a manner similar to that reported by Poole (1969).

Pearcy (1967) estimated the survival from year-class 1 to 2 as about 40%. Pearcy's estimate was considered too high to be used in the model for the following reasons.

1. This survival rate is higher than Poole's estimate of the survival of adult fish.

2. It implies that the simulated population would be growing at an astronomical rate, more than tripling in the next year. Catch statistics for the Southern New England winter flounder do not indicate that the abundance of the species was increasing during the period when Saila and Pearcy conducted their pertinent research.

Therefore, the survival rate of yearling winter flounder was calculated to effect an equilibrium situation on the simulated population. Let

s_1 = annual survival of yearlings

s_2 = annual survival of fish 2 and greater

A = number of fish age 2 and greater

Then

$$YC(2) - s_1 \cdot YC(1)$$

$$YC(3) = s_2 \cdot YC(2) = s_2 s_1 \cdot YC(1)$$

$$YC(4) = s_2 \cdot YC(3) = s_2^2 \cdot s_1 \cdot YC2$$

$$\vdots$$

$$YC(j) = s_2^{j-2} \cdot s_1 \cdot YC(1) \qquad (7)$$

and

$$A = \sum_{j=2}^{12} s_1 \cdot s_2^{i-2} \cdot YC(1) = s_1 \cdot YC(1) \cdot \sum_{j=0}^{10} s_2^j \quad (8)$$

Thus

$$s_1 = A \,/\, YC(1) \cdot \sum_{j=0}^{10} s_2^j \quad (9)$$

If $s_2 = 0.33$, $YC(1) = 83724$ and $A = 18,158$, then $s_1 = 0.1454$ or an instantaneous mortality rate of 1.928. Based on the above discussion of mortality rates, the initial age-class structure shown in Table 3 was assumed for the simulated population.

Tables 2 and 3 were used to estimate the mean fecundity of mature females as 350,296 eggs. Assuming 5996 mature fish (Table 3), 70% females and a 10% successful hatching rate (Saila, 1961b), then

$$(0.7)(5996)(350,296)(.1) = 1.469 \times 10^8$$

hatched eggs result in 83,724 yearling winter flounder or about 55/100,000 if the simulated population is at equilibrium. The agreement between the above and Pearcy's (1962) estimate of 44 yearlings per 100,000 hatched eggs is encouraging, since these results were obtained independently.

In order to apply the model described in Figure 10, a relationship between egg production and recruitment to year-class 1 is required. The ratio of 55 yearlings per 100,000 hatched eggs could be used to define such a relationship as follows:

$$YC_i(1) = 0.000055\, P_{i-1} \quad (10)$$

where i refers to the year. This relationship is density independent in that the survival rate to a year of age is independent of the number of eggs produced. Since the population cannot compensate for loss due to entrainment, even a very low value of M_{p1} will result in an exponential decay of the population eventually approaching zero. This situation is unrealistic since most populations are stabilized by density dependent stock-recruitment relationships (Ricker, 1954). Ricker (1958) suggested the following density dependent relationship between egg production and recruitment:

Table 3. Initial age-class structure for the Niantic River winter flounder population.

Age	Number
1	83,724
2	12,172
3	4,017
4	1,326
5	437
6	144
7	48
8	16
9	5
10	2
11	1
12	-
Total	101,892

$$YC_i(1) = P_{i-1} \cdot e^{a-b} \cdot P_{i-1} \tag{11}$$

where a and b are parameters.

Using several values of $YC_i(1)$ and P_{i-1}, a and b can be estimated by simple linear regression after some manipulation of Equation (7). Unfortunately such data is rarely available and when it is, the range of P_{i-1} is often so limited that little confidence can be placed in the results. Therefore a deterministic procedure for parameterizing Equation (11) based on a limited amount of data was developed.

If S_r is the fraction of females in the population, FEC is the average fecundity and R_1 is the expected number of recruits from N_1 mature fish, then using Equation (11),

$$a = \log_e \frac{R_1}{S_r \cdot FEC \cdot N_1} + b \cdot S_r \cdot FEC \cdot N_1 \tag{12}$$

The Ricker stock-recruitment function is based on the assumption that, in the absence of fishing, the population will adjust itself to a maximum level, N_{max}, at which it is in equilibrium (ignoring environmental fluctuations). This unfished equilibrium level occurs when the loss due to natural mortality equals the addition of recruits. Therefore the second parameter of the stock recruitment function can be calculated, if the size and average fecundity of the virgin (unfished) stock and the level of natural mortality are known.

The annual natural mortality rate (M) times N_{max} was set equal to the recruitment to year-class 1 (for N_{max} mature fish) times the survival rate from age 1 to sexual maturity (s). Equation 12 was substituted and the result was solved by b yielding the following:

$$b = \frac{\log_e \frac{M}{s \cdot S_r \cdot FEC} - \log_e \frac{R_1}{N_1 \cdot S_r \cdot FEC}}{FEC \cdot S_r \cdot (N_1 - N_{max})} \tag{13}$$

A reasonable estimate of N_{max} was sought in the historical literature. The longest series of relative abundances for a New England winter flounder fishery was recorded by the United States Fish and Wildlife Service Hatchery at Boothbay, Maine, from 1910-1940 (reported by Perlmutter, 1947). During this period the Fish and Wildlife Service used fyke nets to collect mature female winter flounder for hatchery use.

This data indicates that the Maine fishery has gone through three rather distinct levels of relative abundance (an indication of population size). The mean catch per net for the periods 1910-1919, 1920-1933, 1933-1940 was 267, 107 and 30 fish respectively. The level of exploitation during the first period was probably very low, since according to the limited amount of catch data available, the level of landings prior to 1919 was low. A comparison of the mean relative abundance between the first and third period indicates that the level of exploitation reached during the late 1930's in Maine waters could reduce the population size of a virgin winter flounder stock by a factor of 7. Perlmutter's (1947) comparison of fisheries throughout New England and New York indicates that later trends in the Maine fishery were reflected in Connecticut waters. Poole (1969) showed that the level of exploitation reached during the late 1930's for the Great South Bay, New York, winter flounder population was similar to the present level of exploitation. There fore, the assumption that $N_{max} = 8N_1$ seems reasonable, since the population size of Boothbay stock may have declined slightly from its virgin level prior to 1910.

Values of other constants in Equations 8 and 9 were assumed, based on the earlier discussion of population parameters (N_1 = 5996, R_1 = 83724, FEC = 350,296, S_r = 0.7, M = 0.33 and $s = s_1s_2$ = 0.05). The estimate of a and b obtained were -9.66 and 7.54 x 10^{-9} respectively. These results are based on the assumption that the average fecundity of the breeding stock was similar for populations of N_1 and N_{max} individuals. When the population was simulated using these estimates for zero fishing mortality, it converged to an equilibrium population level of 47,842 mature fish. This level is about 8 times the initial level assumed for the simulated population.

The life expectancy of a nuclear power plant is about 35 years. Therefore, the system described above was simulated for 35 years at various levels of entrainment mortality (M_{pl}) and for an additional 65 years with M_{pl} = 0. The results are shown in Figure 11.

Since the Ricker stock-recruitment function was based on a rather limited amount of information, the model was also run under the unrealistic assumption that the recruitment is density independent (Equation 10) and thus there is no compensation for entrainment losses. The results do not differ drastically from Figure 11 during the first 35 years of the simulation. The population levels reached in 35 years are 4 to 40 percent lower than

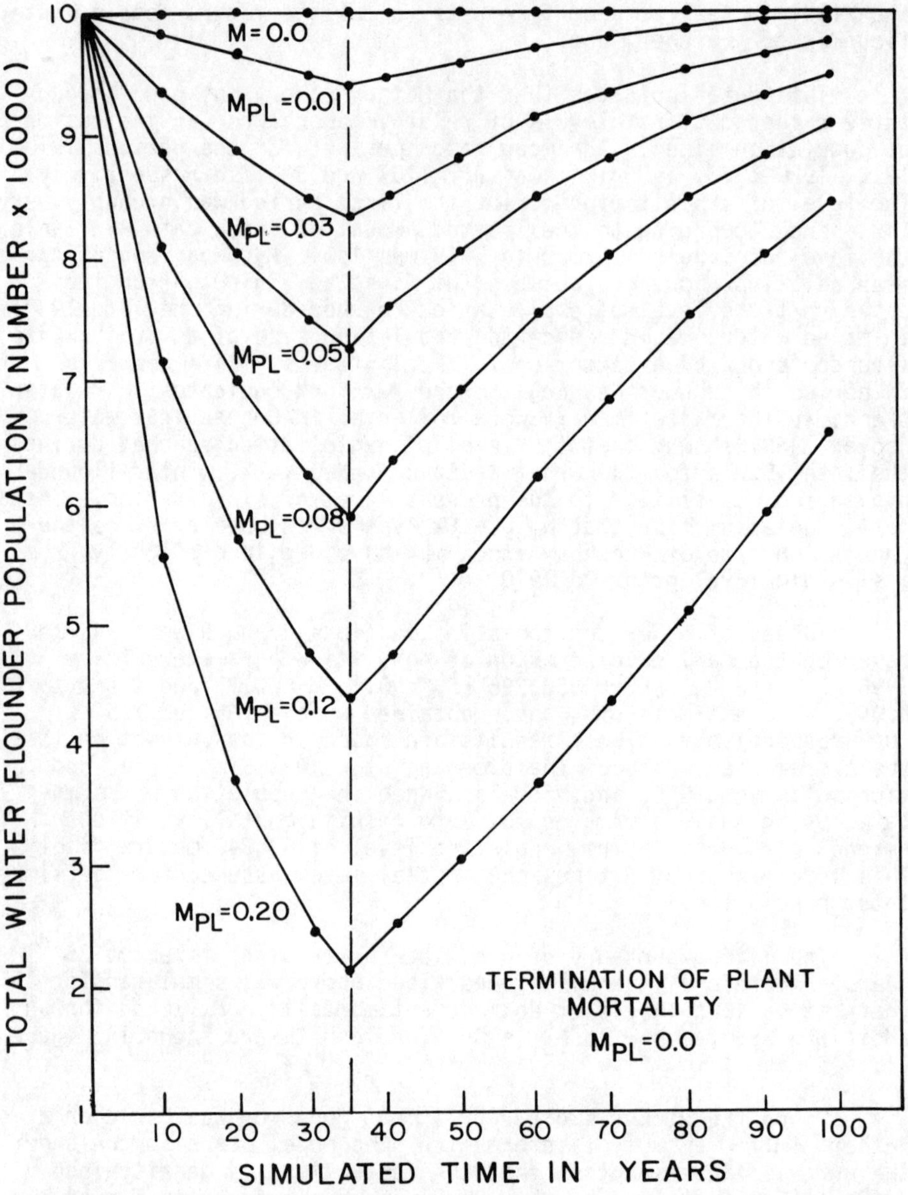

Figure 11. Simulated total population of winter flounder breeding in the Niantic River with various levels of entrainment mortality (M_{pl}).

with the Ricker stock-recruitment function with the largest difference occurring for $M_{pl} = 0.20$. Using Equation 10, there is no recovery of the population after entrainment is terminated. The results based on Equation 10 are conservative since they do not permit compensation for entrainment mortality and thus they are estimates of the potential effect of the plant.

For the level of M_{pl} indicated by the larval dispersion model (1% or less), Figure 11 indicates a potential impact of a 6% reduction in total population level after 35 years. If all compensatory mechanisms are ignored, a 9% reduction is indicated.

It must be emphasized that these results probably overestimate the effect of entrainment since they ignore the survival of some organisms passing through the plant, immigration of winter flounder, input of larvae from outside the bight, density dependent growth, fecundity, adult mortality and in some cases larval mortality, and a reduction in fishing mortality as fishermen become discouraged and reduce their effort. Therefore, the results should not be taken as predictions, but as limits on the potential danger from various levels of M_{pl} based on the available information.

DISCUSSION

This paper reports on a continuing effort to assess the impact of entrainment of winter flounder larvae. Some aspects of the technique may be modified to simulate the system more realistically as information becomes available. The results are being verified against hydrographic and biological data collected in the region.

The model of larval transport has been advanced using the assumption that no significant vertical stratification in the distribution of the organisms exists. There is, however, some built-in flexibility in the model which would allow vertical variations to be incorporated. The transport equation (4) can be modified by adjusting the coefficients of the convective terms, which arise from the vertical integration of the product of velocity and concentration, and which are unity if either the velocity or concentration is uniform over depth. A number of profiles have been examined and for most cases the coefficients are only a few percent less than unity. This situation results from a nearly vertically uniform flow rate that is characteristic of a tidally dominated system. If the model is modified to consider vertical stratification, some increase in the retention of larvae in the Niantic River is likely, but the overall effect on the magnitude of M_{pl} is expected to be minor.

A number of verification studies are in process. Predicted currents have been successfully verified against current meter records. A tag-and-recapture program is underway to estimate the number of adult winter flounder which spawn in the Niantic area. With certain assumptions the number of hatched larvae can be calculated and used to predict, using the larval transport model, the temporal and spatial distribution of organisms for several weeks after the initial hatching. Predicted concentrations will then be compared to those obtained from an ichthyoplankton sampling program conducted at several stations around the bay. The number of larvae captured at the intakes can also be used for verification. The sampling program will also yield information about the vertical distribution of organisms.

Although the amount of data required for complete verification is considerable, it could not be used to estimate M_{pl} without the models developed during this study. For example, it is possible for several percent of the larvae spawned in the Niantic River to be entrained, with no effect on the number of future recruits. This situation could occur if all entrained organisms were destined to be transported out of the system into Long Island Sound.

Only the direct effect of entrainment on a single species has been considered during this study. The results are still subject to further verification. Nevertheless, the integrated simulations of biological and hydrodynamic systems reported herein suggest that the potential impact of entrainment on the winter flounder of the Niantic River is less than a 1 % reduction in year-class strength with a 6% reduction in population size after 35 years of plant operation.

ACKNOWLEDGMENTS

This study was supported by research funds provided by the Northeast Utilities Service Corporation, Hartford, Connecticut.

LITERATURE CITED

BERRY, R. J., S. B. SAILA and D. B. HORTON. 1965. Growth studies of winter flounder, Pseudopleuronectes americanus (Walbaum), in Rhode Island. Trans. Am. Fish. Soc. 94(3): 259-264.

ELDER, J. W. 1959. The dispersion of a marked fluid in turbulent shear flow. J. Fluid Mech. 5: 544-560.

LABELL, M. J. 1939. A biological survey of the salt waters of Long Island, 1938. Report on certain fishes. Winter flounder (Pseudopleuronectes americanus). Suppl. 28th Ann. Rept., N.Y. Cons. Dept., Pt. 1: 63-96.

LEENDERTSE, J. J. 1967. Aspects of a computational model for long-period water waves propagation. Rand Co., Santa Monica. Memo RM 5294-PR.

LEENDERTSE, J. J. 1970. A water-quality simulation model for well-mixed estuaries and coastal seas, volume 1. Rand Co., Santa Monica. Memo RM-6230-RC.

LUX, F. E. 1969. Length-weight relationship of six New England flat fishes. Trans. Am. Fish. Soc. 98: 617-621.

PEARCY, W. G. 1962. Ecology of an estuarine population of winter flounder, Pseudopleuronectes americanus. Bull. Bingh. Ocg. Coll. 18(1): 1-78.

PERLMUTTER, A. 1947. The blackback flounder and its fishery in New England and New York. Bull. Bingh. Ocg. Coll. 11(2): 1-92.

POOLE, J. C. 1969. A study of winter flounder mortality rates in Great South Bay, New York. Trans. Am. Fish. Soc. 98(4): 611-616.

RICKER, W. E. 1954. Stock and Recruitment. J. Fish. Res. Bd. Canada 11(5): 559-623.

RICKER, W. E. 1958. Handbook of computation for biological statistics of fish populations. J. Fish. Res. Bd. Canada 119: 1-300.

SAILA, S. B. 1961a. A study of winter flounder movements. Limn. and Oceanog. 6(3): 292-298.

SAILA, S. B. 1961b. The contribution of estuaries to the offshore winter flounder fisheries in Rhode Island. Gulf and Caribbean Fisheries Inst. Proceedings 14th Annual Session, November 1961, p. 95-109.

SOME COMMENTS ON THE THERMAL EFFECTS OF POWER PLANTS ON FISH EGGS AND LARVAE

J. R. Schubel
Marine Sciences Research Center
The State University of New York at Stony Brook
Stony Brook, L.I., New York 11794

ABSTRACT

Fish eggs and larvae entrained by steam electric stations with once-through cooling systems are exposed to an initial rapid rise in temperature which is approximately equivalent to the temperature rise across the condensers. They are exposed to the maximum ΔT during passage through the plant and to the point of discharge and then to decreasing excess temperatures (i.e., temperatures above those to which they have been acclimated) as they are carried down the plume. The actual time-temperature exposure histories depend primarily upon plant design and the characteristics of the receiving waters. Site studies have failed to establish unequivocally the cumulative effects of the several stresses experienced by entrained fish eggs and larvae, and separation of the individual effects of these stresses has rarely, if ever, been possible. Most laboratory studies of the thermal requirements of fish eggs and larvae are also of little use in predicting the thermal effects of entrainment because of the inappropriateness of the time-temperature histories. The few laboratory studies of fish eggs and larvae that have been specifically directed at entrainment questions have shown clearly that the thermal effects of entrainment can be assessed in the laboratory. These few data are reviewed. Blueback herring, alewife, American shad, and striped bass eggs were able to survive exposure to typical time-excess temperature histories with ΔT's of at least 10°C above the average temperature on the spawning grounds without any harmful effects. Larvae were generally able to withstand exposure to ΔT's of 12°C above acclimation temperatures of 5° to 15°C for at least 40 minutes without any significant increases in mortality. At higher excess temperatures, acclimation temperature and duration of exposure became increasingly more important.

INTRODUCTION

Although there has been a vast amount of research on the effects of temperature on fish and other aquatic organisms (see, for example, the bibliographies of Coutant, 1968, 1969, 1970a, 1971a; Kennedy and Mihursky, 1967; Raney *et al.*, 1973), there has been considerably less research on the effects of temperature on fish eggs and larvae, and very little of the published work is of direct use in predicting the thermal effects of power plants on entrained fish eggs and larvae. In nearly all of the published laboratory thermal studies of fish eggs and larvae, the organisms have been subjected to some constant elevated temperature for all, or nearly all, of the experimental period (e.g., Edsall, 1970; Shannon, 1970; Bradford et al., 1968; Kelley, 1968; Albrecht, 1964; Price, 1940; and many others). Such exposure histories are not representative of those experienced by planktonic organisms entrained in the once-through cooling system of a steam electric station. Steam electric stations with once-through cooling systems remove large volumes of water from the environment and return them at elevated temperatures. Organisms smaller than the mesh size of the intake screens--generally 9 to 13 mm--are carried along with this water.

Planktonic organisms, including fish eggs and larvae, may be drawn into the cooling water intake of a power plant with a once-through cooling system, passed through the condensers and discharged back into the environment; or they may be entrained into the thermal plume along with the diluting water without passing through the plant. The organisms that pass through the plant are subjected initially to a very rapid rise in temperature, approximately equivalent to the temperature rise across the condensers. They are exposed to this maximum excess temperature (i.e., the temperature above the normal ambient, expressed as ΔT) during passage through the plant and to the point of discharge, and then to decreasing excess temperatures as they are carried down the plume. In the near field, the rate of cooling along the plume is determined almost entirely by the intensity of the lateral and vertical mixing. Planktonic organisms that are entrained into the thermal plume without having passed through the plant are subjected to a less rapid rise in temperature and to somewhat lower excess temperatures. The actual time-excess temperature histories experienced by planktonic organisms will vary from plant to plant, and to a much smaller extent, with time at any given plant. The time-excess temperature exposure history depends upon the design of a plant's condenser system, its intake-discharge system, the characteristics of the receiving waters, and the plant's operating load.

The extent of the region of excess temperature and the distribution of excess temperature in the waters contiguous to a plant are determined by the volume of cooling water pumped, the fraction of the available water used for cooling, the size and design of the plant, the type and placement of the discharge structure, and the characteristics of the receiving waters.

The primary purpose of this report is to assess the thermal effects of steam electric stations with once-through cooling systems on entrained fish eggs and larvae. The assessment is made largely from my own data, from the available literature that is directly pertinent, i.e., those laboratory studies in which appropriate time-temperature histories have been used, and from field studies at operating plants.

Some Characteristics of Steam Electric Stations

Typically a steam electric station (SES) consists of a heat source, boiler, turbine, generator, and a condenser system (Fig. 1). Steam from the boiler drives the turbine which spins the generator to produce electricity. In order to repeat the entire cycle efficiently, the steam must be condensed. The most economical way of achieving this is by passing cooling water from the environment through the condenser system and back into the environment at an elevated temperature--a once-through cooling system. The heat exchange takes place in the condenser units. Each condenser unit consists of a bank of up to 10,000 tubes, each approximately 2.5 cm. in diameter and from 10 to 25 m. long. The cooling water passes through the insides of the tubes while steam from the turbine exhaust passes around the outsides. Since the flow speed of the cooling water through the condensers is generally about 1.5 to 2.5 m./sec^{-1}, heat is added to the cooling water for only about 5 to 20 seconds.

Typical Time-Excess Temperature Histories

There have been several summaries of the maximum ΔT's experienced by organisms entrained by power plants. The Water Resources Council (1968) reported that the average temperature rise across the condensers of U.S. power plants was 8.3°C (15°F). According to Coutant's (1970b) examination of 61 power-reactor designs filed with the U.S. Atomic Energy Commission, however, the predicted average ΔT at these plants would be 10.8°C (19.4°F), and the range from 5.6°C (10°F) to 18.6°C (34°F). European investigators have documented excess temperatures of 7.5°C (13.5°F) and 11°C (19.8°F) at fresh water plants, and 12° to 16°C (21.6° to 38.8°F) at salt water plants (European Inland Fisheries Advisory Commission, 1969). Excess temperatures much less than 7.7°C (12°F) are, according to Lee (1970), generally impractical because of the economics of pumping cost and equipment size. Thermal death is clearly a dose response, although many investigators have failed to recognize this. Total passage time through a plant is only of the order of a few

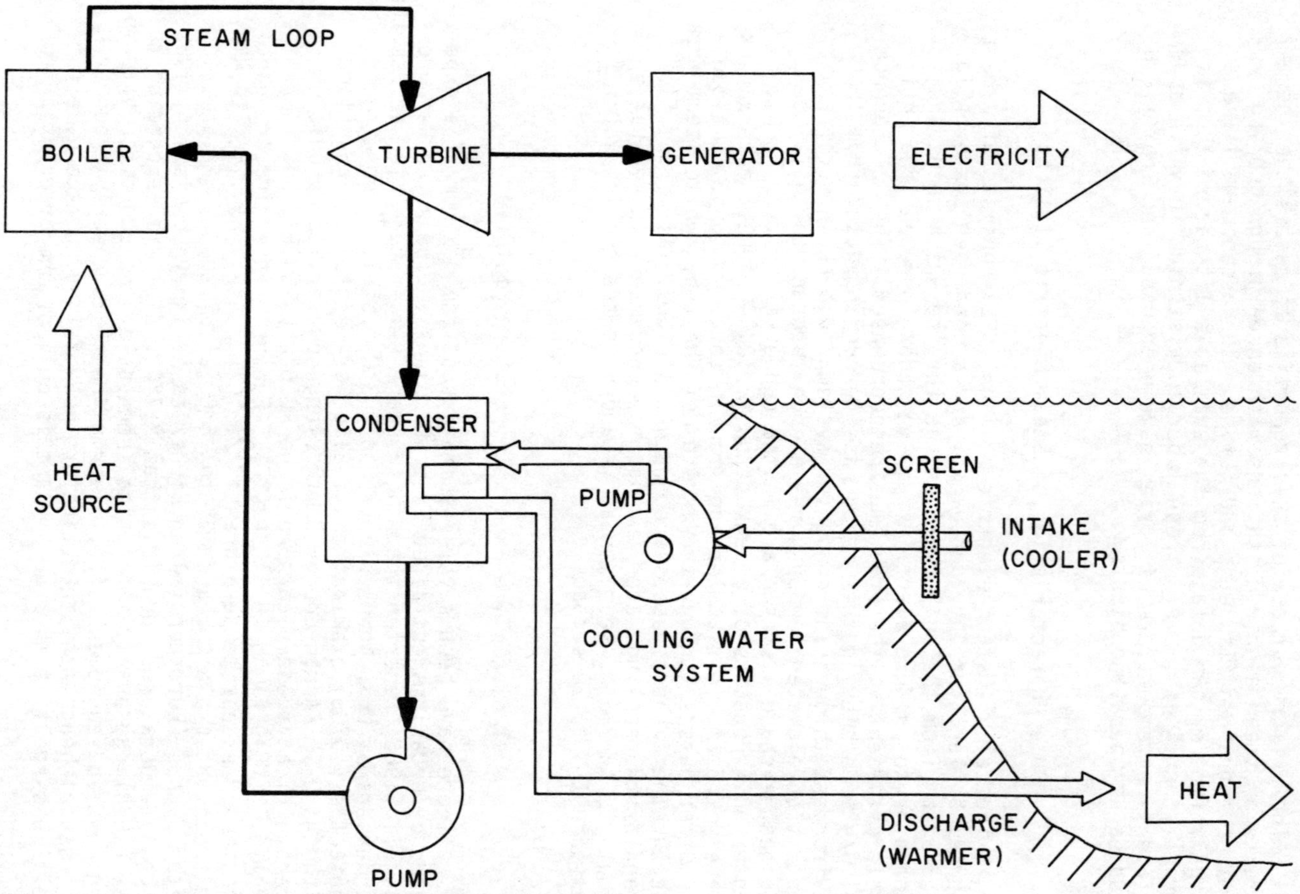

Figure 1. Schematic of a steam electric station.

minutes, but the exposure time to nearly a full ΔT may be extended to an hour or more at plants with long discharge canals.

Nuclear and fossil-fueled steam stations are similar in their operation. The primary difference between them is the energy source. The cooling systems are alike in essentials, although nuclear plants use larger quantities of cooling water per unit of power, generally about 1.5 times as much, because lower steam pressures are used with nuclear fuel. The lower pressure results in greater heat rejection and therefore necessitates larger volumes of cooling water to maintain a desired ΔT across the condensers. According to the U.S. Atomic Energy Commission (1969), nuclear power plants, planned or in operation, average about 3 m^3 of cooling water per minute per megawatt (electrical).

A steam electric station with a once-through cooling system may discharge the cooling water through a canal or a diffuser, or as a high speed surface or submerged jet. The most rapid dilution of heated effluent, and therefore the most rapid reduction in temperature, is attained with long diffusers that discharge the water through a large number of small ports. Submerged jets are generally the next most effective mode of discharge in terms of temperature reduction, and are considerably less expensive to build than diffusers. Canals are the least desirable type of discharge from the standpoint of thermal effects. Relatively little cooling takes place in canals unless cool diluting water is added to the canal close to the condenser system with auxiliary pumps. Canals commonly aggravate the thermal effects of entrainment, and can have a number of other serious drawbacks (Coutant, 1970b).

Entrained organisms are subjected to several other stresses in addition to excess temperatures. The pressure change during passage through a plant is equivalent to about 4 to 9 m. of head (Adams, 1969) and turbulence levels are high in the intake pumps, in the condenser waterbox, and frequently at the point of discharge. In addition, at many stations, a biocide is added intermittently in front of the circulating pumps to control the growth of organisms in the condenser tubes.

The maximum ΔT is not the only important factor. The time-temperature exposure histories are of critical importance in determining the thermal effects of entrainment. The time-temperature exposure histories associated with a particular plant depend upon the design of the plant--particularly its condenser and discharge systems--on its operating load, and on the character of the receiving waters.

Adams (1969) presented a detailed time-temperature history for passage through the Pittsburgh power plant of the Pacific Gas and Electric Company. Pritchard and Carter (1972) have developed, for a variety of power plant operating and design criteria, time-excess temperature relationships that give the maximum time of exposure of entrained organisms to any specified ΔT. Each relationship is based on a specific maximum ΔT, corresponding to the temperature rise across the condensers, and on a particular set of cooling conditions ranging from surface cooling only to surface cooling with various degrees of dilution. The rate of dilution is dependent upon a variety of factors including the speed of the discharge at the orifice, the width of the discharge orifice and the circulation of the receiving waters. Using data from Pritchard and Carter (1972), Schubel (1974) presented a number of time-excess temperature histories characteristic of a variety of plant designs. Coutant (1970b, Fig. 3) presented a hypothetical time-excess temperature curve for a plant with a diffuser discharge and a curve for a plant with a discharge canal (Fig. 2a). A time-excess temperature history "typical" of a plant with a jet discharge is presented in Figure 2b.

Thermal Effects of Entrainment on Eggs

Laboratory Studies

A small number of pertinent laboratory studies have been conducted with fish eggs (Schubel, 1974; Schubel and Auld, 1972a, b, 1973a, b; and Frank, 1973). Frank (1973) studied the effects of exposure of carp eggs in different stages of development to 10-minute periods of acute thermal shock. Schubel and Auld (1972a, b, 1973a, b) presented results of preliminary laboratory experiments in which American shad (*Alosa sapidissima*), alewife (*Alosa pseudoharengus*), blueback herring (*Alosa aestivalis*) and striped bass (*Morone saxatilis*) eggs were subjected to time-temperature histories tyipcal of those experienced by organisms entrained in the once-through cooling systems of power plants in Maryland on the Chesapeake Bay. A number of other laboratory studies have reportedly been conducted by scientists for the power industry, but the results are proprietary and have not been released.

One of the most comprehensive laboratory studies of the thermal effects of entrainment on fish eggs is that of Schubel (1974). Schubel subjected blueback herring, alewife, American shad, white perch, and striped bass eggs to time-temperature histories typical of those that would be experienced by organisms entrained by power plants with a variety of designs and operating conditions. The eggs were quickly immersed into water of the temperature appropriate

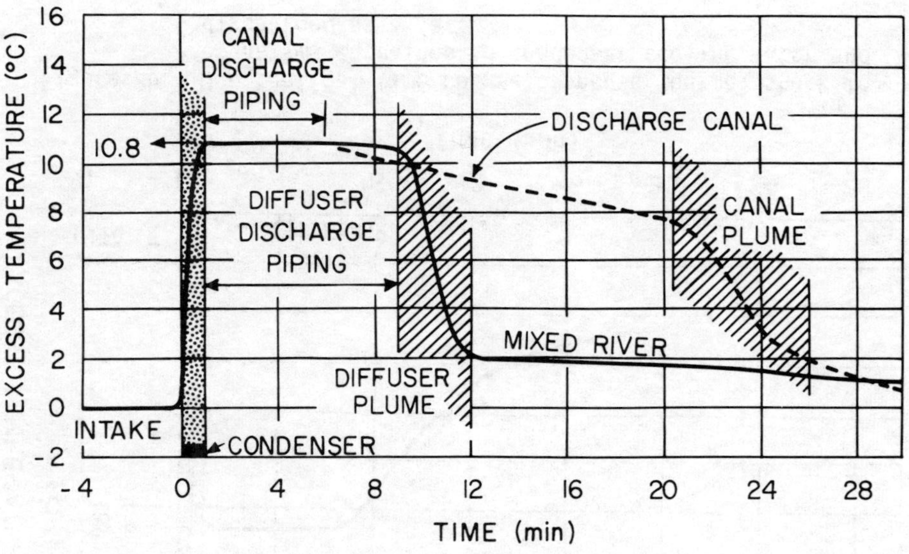

Figure 2a. Hypothetical time-courses of acute thermal shock to organisms entrained in condenser cooling water and discharged by diffuser or via a discharge canal (after Coutant, 1970b).

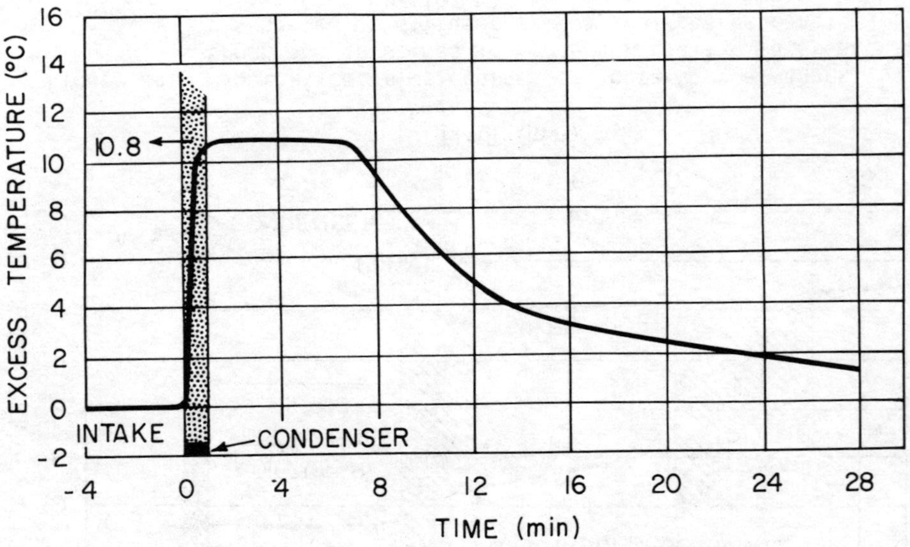

Figure 2b. Hypothetical time-course of acute thermal shock to organisms entrained in condenser cooling water and discharged as a jet.

to produce a desired ΔT, then cooled at a pre-determined rate. The maximum excess temperatures (ΔT) tested ranged from 6° to 10°C and the time of exposure to a maximum ΔT from 2.5 to 60 minutes. The period of cooling to the final temperature, 1°C above the base temperature, ranged from 60 to 300 minutes. Schubel's (1974) time-excess temperature histories are shown in Figure 3. All excess temperatures were keyed to the average temperature on the spawning grounds at the time the fish were collected--the base temperature. The final experimental temperature, 1°C above base, is analogous to the far-field temperature that might be observed in nature.

For each species, six to eight experiments were run with eggs in different stages of development. Each experiment involved six to twelve samples of 200 to 400 eggs taken from a single female and fertilized with the milt from several males. In each experiment, a control and four to six time-excess temperature histories were tested. The control treatment consisted of maintaining the eggs at a constant temperature (the base temperature) throughout the experiment. These eggs were subjected to the same handling given to the other samples. Each of the samples of experimental eggs was exposed to the appropriate time-temperature history, then incubated in a constant temperature bath set at 1°C above the base temperature until hatching occurred. The experiments were terminated shortly after hatching. Calculations were made of the absolute hatching successes--the percentage of eggs in the control and in each experimental sub-sample that hatched successfully. The larvae were examined under the microscope for any gross morphological abnormalities.

For each species, experiments were run with eggs in several different stages of development. The stages ranged from early cleavage through late embryo, except in the case of striped bass where only embryonic stages were tested. Experiments were attempted with striped bass eggs in earlier stages of development, but the mortalities in all of the sub-samples, including the controls, were so high--greater than 95%--that the results were not significant. Schubel (1974) pointed out that further experiments should be conducted with eggs during cleavage stages, but cautioned that natural mortalities of striped bass eggs are so high during the early stages of development that the probability of success is very low.

Schubel's (1974) results for blueback herring, alewife, American shad, and striped bass are summarized in Table 1. Only two experiments were conducted with white perch and they have not been included. Schubel's statistical analysis showed, for each of the five species he tested, that the differences in hatching success among the various time-temperature treatments--including the

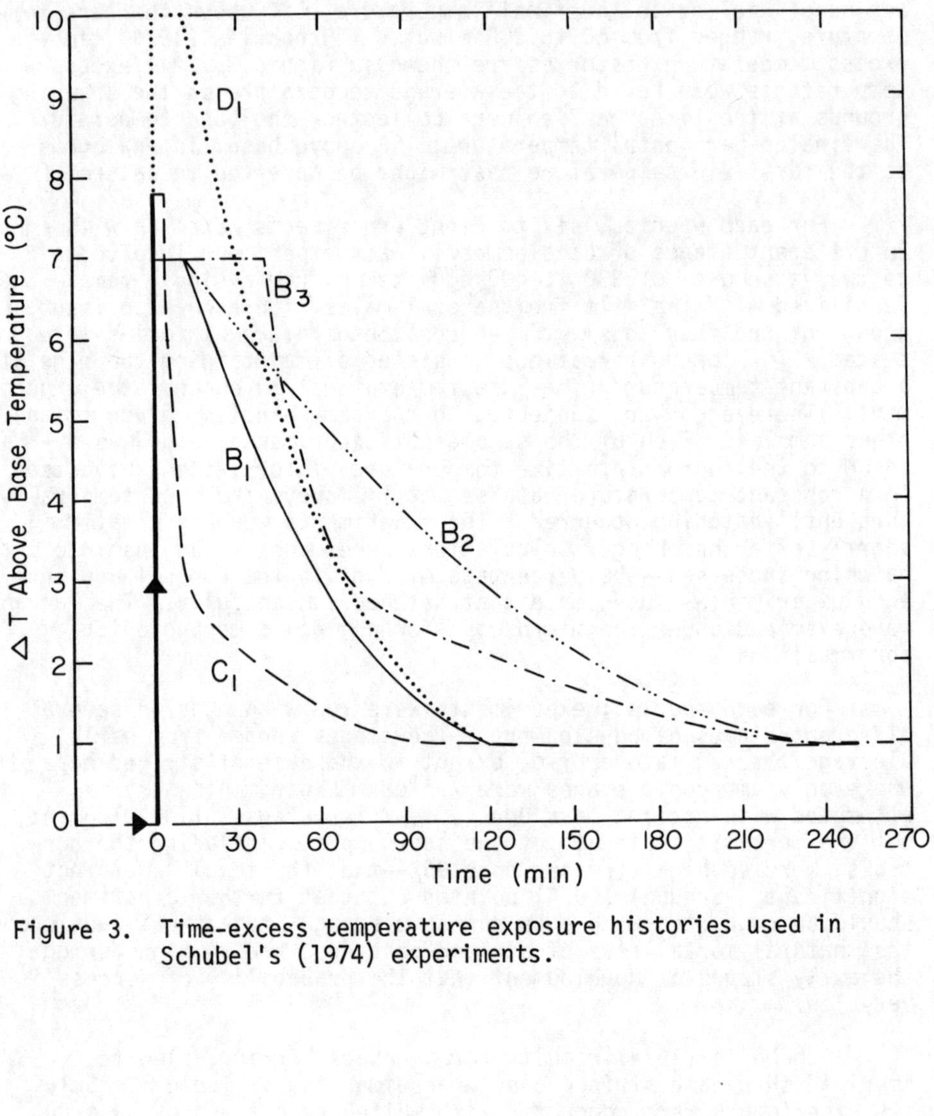

Figure 3. Time-excess temperature exposure histories used in Schubel's (1974) experiments.

Table 1. Summary of Schubel's (1974) results for blueback herring, alewife, American shad, and striped bass. In each of Schubel's experiments a control sample of eggs was held at a constant base temperature throughout the incubation period, and experimental samples were exposed to various time-excess temperature histories (Fig. 3) before incubation. Schubel's data for each species have been averaged. The mean and standard deviation of the hatching success of the controls and experimental treatments are given.

Species	Stages of Development Tested	Range of Base Temperature Tested (°C)	Mean (and Standard Deviation) of Hatching Success (%)					
			Control	Time-Temperature Histories				
				B_1	B_2	B_3	C_1	D_1
Blueback herring	Late blastula - late embryo	15.0-18.3	90(15)	88(12)	90(10)	88(7)	90(9)	90(9)
Alewife	4 and 8 cell - late embryo	12.0-14.5	81(13)	80(14)	78(12)			81(12)
American	Morula - late embryo	16.5	93(4)	90(4)	92(6)	93(4)	92(3)	93(4)
Striped bass	Early embryo - late embryo	14.5-18.3	99(2)	98(2)	99(1)	98(2)	98(2)	99(1)

time-constant temperature control treatment--were not significant at the 1% level, and that the mean hatching success of the controls did not differ significantly from the mean hatching successes associated with any of the experimental time-excess temperature treatments. Schubel did find, however, that fertilized eggs from some females had higher hatching success than others, except in the case of striped bass.

Schubel's (1974) data did not indicate any differences in the thermal sensitivity of eggs in different stages of development. Other investigators (e.g., Kelley, 1968; Combs, 1965; Volodin, 1960; Swarup, 1958, 1959a,b; Hayes and Armstrong, 1942) have reported that eggs do differ in their thermal sensitivity in different stages of development, being especially sensitive during early cleavage stages. Shannon (1970) reported that freshly fertilized striped bass eggs were more susceptible to thermal shock than eggs that had been incubated for 16 to 44 hours at a lower temperature.

Some of the common deformities that have been related to temperature include spinal deformities (kyphosis, lordosis, scoliosis, vertebrae fusion), and edema (Gunter and Ward, 1961; Volodin, 1960). Swarup (1958; 1959a,b) subjected freshly fertilized eggs of the fresh water fish, Gasterosteus aculeatus to 5 minute periods of hot shock. Eggs were also subjected to 1-1/2 hour to 3 hour periods of cold shock.

The temperatures for "heat treatment" ranged from 32.5° to 35.5°C, and temperature of the "cold treatment" was 0°C. Since Swarup did not report a base temperature, the ΔT's are not apparent. Swarup reported that abnormalities increased with the severity of the temperature treatment. Abnormalities were observed during cleavage and gastrulation. Cleavage abnormalities included unequal division of the blastodisc and inhibition of cleavage, while gastrulation abnormalities included formation of only the anterior part of the head, formation of spina bifida, and formation of only trunk and tail-like structures. Abnormalities were produced by both hot and cold treatments, although the frequency of abnormalities was greater with increased temperature. All of the investigators except Swarup exposed eggs to some constant temperature for extended periods--hours or even days. Apparently, the thermal doses used in Schubel's (1974) experiments were all below the critical threshold. Schubel reported that microscopic examination of exposed eggs and larvae "failed to produce any morphological or behavioral evidence of aberrant development that could be attributed to exposure to the various time-excess temperature histories." Schubel's experiments, however, were designed to assess gross thermal effects, effects that would be manifested by differences in hatching success.

Schubel's microscopic examinations were limited, and much more detailed observations are required to establish unequivocally whether or not any subtle genetic effects are produced by exposure to time-temperature histories typically experienced by organisms entrained by steam electric stations. An observational method that might be useful is time-lapse cinephotomicrography.

Schubel and Koo (in preparation) conducted additional experiments in which striped bass, blueback herring, and American shad were subjected to time-excess temperature histories similar to those depicted in Figure 3, and to higher ΔT's--15° and 20°C. All excess temperatures were keyed to the average surface water temperature at the time of spawning. Analysis of these data is incomplete, but preliminary inspection indicates that a ΔT of 20°C results in virtually total mortality within 1 to 2 minutes of the initial exposure. Unfortunately, no detailed observations of egg development following exposure to time-excess temperature histories were made.

Field Studies

There are, in this investigator's opinion, no published site studies that are of any significant value in assessing the thermal effects of entrainment on fish egg survival and development.

Thermal Effects of Entrainment on Larval Fishes

Laboratory Studies

The only laboratory studies the author is aware of in which larval fishes have been subjected to time-temperature histories characteristic of those experienced during entrainment at a steam electric station are those of Hoss *et al.* (1973), and Schubel and Koo (in preparation). Hoss *et al.* (1973) exposed larvae of Atlantic menhaden (*Brevoortia tyrannua*), spot (*Leiostomus xanthurus*), pinfish (*Lagodon rhomboides*), and three species of flounder (*Paralichthys* spp.) to excess temperatures of 12°, 15°, and 18°C for 0, 10, 20, 30, and 40 minute periods. Larvae were acclimated at 5°, 10°, 15°, and 20°C, shocked to a higher temperature, held at that temperature for a pre-determined time, then returned immediately to the acclimation temperature. Although the rate of cooling is more severe than that observed at operating plants, the method is convenient since it is easily repeated. Furthermore, it does provide very useful results since they are "environmentally conservative." Hoss *et al.* (1973) reported survival rates for exposed fish one day and seven days after exposure. In most cases, survival after seven days was essentially the same as after one day. In those cases where it was not, mortalities were also higher in the controls indicating that some factor other than temperature was involved.

Hoss *et al.* (1973) reported that, except for Atlantic menhaden, survival was not significantly affected by exposure to a ΔT of 12°C with any of the acclimation temperatures and exposure times tested. Atlantic menhaden acclimated to 10°C could generally survive an excess temperature of 12°C, but survival of fish acclimated to higher temperatures was significantly decreased following exposure to this ΔT. Survival after a 15°C shock was, for all species tested, clearly dependent upon acclimation temperature and the duration of exposure; percent survival increased with decreasing acclimation temperature and decreasing exposure period. Exposure of all species investigated by Hoss *et al.* (1973) to an excess temperature of 18°C resulted in significantly increased mortalities for all acclimation temperatures and exposure periods, but both of these factors clearly had an effect on the percent mortalities. For example, over 50% of the spot larvae acclimated to 5°C could survive an 18°C shock regardless of exposure time, while spot "acclimated to 10°C showed greatly reduced survival to the same shock when exposure was greater than 10 minutes . . ." and spot "larvae acclimated to 15°C could not survive an 18°C shock for any of the exposure periods tested."

Hoss *et al.* (1973) made observations of the critical thermal maximum (CTM), oxygen consumption, and behavior of larval fish exposed to different time-temperature histories. Hoss *et al.* did not find any persistent effects of thermal shock either on oxygen consumption or growth rate. This author agrees with Hoss *et al.* that CTM data are of little value in predicting the thermal effects of entrainment by power plants on fish larvae.

Their behavioral observations are also very pertinent. Exposure of larval fish to an excess temperature of 12°C generally did not produce any signs of distress, but exposure to excess temperatures of 15° and 18°C usually caused immediate reaction including loss of equilibrium, erratic swimming, violent jumping and convulsions. If the fish survived this period, they frequently behaved relatively normally for the remainder of the exposure period. Their reactions to the rapid rate of cooling were generally even more violent.

Schubel and Koo (in preparation) exposed striped bass, blueback herring, and American shad larvae to time-excess temperature histories similar to those depicted in Figure 3. In addition, higher ΔT's of 15° and 20°C were also tested. The excess temperatures were keyed to the average surface water temperature at the time of spawning. Analysis of these data is incomplete, but preliminary inspection indicates that a ΔT of 20°C resulted in virtually total mortality within one to two minutes of exposure. A ΔT of 15°C also produced significant increases in mortality, and loss of equilibrium among surviving larvae. Larvae were typically stunned when exposed to a ΔT of 15°C, and sometimes remained motionless for more than an hour after the temperature had decreased to the base temperature. Clearly a shock of 15°C would significantly increase the vulnerability of the larvae of these species to predation.

Experiments with larvae beyond yolk sac absorption are complicated by difficulties in providing the larvae with a suitable diet to avoid the effects of starvation. Hoss et al. (1973) successfully fed larval menhaden, spot, pinfish, and flounder with brine shrimp and dry commercial food. Schiemer and Schubel (1974) constructed a system that provided larvae with a continuous supply of tempered estuary water. Their device was capable of maintaining temperature to within ±0.2°C of a desired temperature while the temperature in the estuary was slowly varying by as much as ±8°C. Their hope was that such a system would supply the larvae with their natural diet, that there would be no effects of starvation, and that the constant temperature would allow assessment of the effects of previous exposure to time-temperature histories typical of those experienced at power plants. While their device was an engineering success, it was a biological failure. Although water was pumped directly from the spawning grounds on the Potomac estuary (Md.)--an area rich in plankton--the effects of larval starvation were readily apparent in their experiments.

Field Studies

There have been numerous site studies of entrained larval and juvenile fish. Edsall and Yocum (1972) cite reports of heavy kills of entrained fish larvae at several power plants. Kerr (1953) passed small (2 to 6 cm long) salmon and striped bass through a condenser of the Contra Costa plant (California), and observed the fish for five to ten days after exposure. Kerr reported a ΔT of about 16°C, and a total exposure time of 3.5 to 5 minutes, but he did not report the base temperatures. In ten tests with a total of 100 salmon, Kerr (1953) found that "these fish withstood the 16 degree temperature rise of the condenser with no fatalities." Kerr concluded that for "yearling king salmon of the size able to pass a 3/8 inch mesh screen . . . . passage through the plant constitutes no great hazard and survival will be extremely high." Kerr's tests with striped bass indicated 94 percent survival of entrained fish at the end of the five-day observation period. Behavioral responses of the thermally shocked fish were not reported, nor were they apparently considered in assessing the effects of entrainment, although Kerr did point out that small striped bass "readily go into a state of shock without apparent reason."

One of the first reported studies on the effects of entrainment on fish larvae was made by Marcy (1971). He reported heavy mortalities of larval fish at the Connecticut Yankee's Haddam Neck Plant on the Connecticut River when temperatures in the plant's 1.83 km long discharge canal exceeded 30°C. The maximum ΔT at the Haddam plant is 12.5°C, and the total travel time to the end of the canal is 50 to 100 minutes. The heavy mortalities were attributed to "heat shock and prolonged exposure to elevated temperature," but Marcy also reported that "the majority of dead specimens were mangled." In later studies at the same plant, Marcy (1973) reported that in June and July nonscreenable fish were abundant near the plant's intake and that about "80% of the mortality in the canal was caused by mechanical damage and 20% was attributable to heat shock and prolonged exposure to temperatures elevated above 28°C." When canal temperatures were between 29° and 33.5°C no entrained fish survived to the end of the canal, and at temperatures greater than 35°C, 100% mortality occurred within the plant itself. These results are in contrast to a preliminary study in which Marcy (1969) reported that larval river herring were able to successfully pass through the condenser when the maximum temperature was 34°C.

Markowski (1959) pumped marked fish through the Roosecote Station, Cavendish Dock (U.K.), to determine the effects of entrainment on young fish. He reported that there was no evidence of damage from temperature or chlorine, but that some of the larger organisms suffered mechanical injury. The exposed organisms were held for several days after exposure to observe any delayed effects, but none occurred. Markowski did not report any time-temperature histories, and his data are largely qualitative. A significant fraction of his observations were made during periods of other than peak summer temperatures.

DISCUSSION AND CONCLUSIONS

Organisms carried through steam electric stations with once-through cooling systems are subjected not only to elevated temperatures, but also to pressure changes, to mechanical agitation, and frequently to a biocide. This sequence of stressful events has sparked the concern of many environmental scientists and has provided them with research funds for a number of years. Still, despite the many millions of dollars that have been spent on "thermal pollution" research, assessments of the cumulative effects of the several stresses experienced during entrainment are equivocal, and separation of the individual effects of these stresses is rarely possible. Development of a capability to predict the "predation" rate of a proposed plant is still lacking, and the capability of predicting the cumulative effects of entrainment by an existing or proposed plant on an entire eco-system appears to be remote.

The vast amounts of money that have been spent on site studies have generally yielded little, other than a chronicling of the fluctuations of the resident populations. Rarely can the effects of a plant be unequivocally established because of the large natural spatial and temporal variations; the signal to noise ratio is too low.

It appears to this investigator that diagnostic data on entrainment must come from appropriate laboratory studies. Few laboratory studies, however, have been conducted that can be used directly to assess the thermal effects of entrainment on fish eggs and larvae. In most of the studies the organisms have been subjected to some constant elevated temperature for extended periods of time; hours or, more commonly, days. Such time-temperature histories are clearly not representative of those experienced by organisms entrained by power plants. A further complication is that a variety of measures of "lethal temperature" have been used, including: (1) the temperature reached before death when heated at a constant rate (e.g., Huntsman and Sparks, 1924), (2) the temperature at which a certain fraction of the exposed organisms survive for a given length of time (e.g., Hathaway, 1927), and (3) the average survival at a given constant temperature. While all of these may be useful measures of lethal temperature, it is difficult to use any of them directly in predicting the effects of exposure to time-temperature histories typical of power plants. Coutant (1970b) has pointed out that if LD_{50} values were available for even two time periods, one could at least connect the two points with a straight line to yield an approximate thermal resistance time curve.

The lack of appropriate laboratory studies has prompted some investigators to turn to field studies for answers to the pertinent questions. According to Adams (1968), "Field experiments seem to offer a much more meaningful evaluation of potential problems from thermal power plants than laboratory experiments." In the author's opinion, however, there is little hope of assessing the individual and cumulative effects of the several stresses experienced during entrainment from field studies. Rather, field data and plant operating conditions should be used to design appropriate, meaningful, laboratory studies. Adams (1968) is correct that, "It is useless to conduct 24 hour TL_m tests on selected organisms to evaluate the effects of condenser passage when the passage time from onset of temperature increase to decline is a matter of minutes rather than hours." Schubel (1974), Schubel and Auld (1972a,b, 1973a,b), and Hoss et al. (1973), however, have clearly shown that diagnostic experiments can be designed to assess the thermal effects of entrainment on fish eggs and larvae by existing plants. These data can also be used to predict the effects of entrainment by proposed plants, and to set appropriate design criteria to minimize thermal effects on entrained organisms. Schubel (1974) pointed out that the conclusions of his egg experiments ". . . should be transferable to power plants in which similar time-temperature histories are experienced by fish eggs. If such plants are important predators of eggs of these species, the predation is apparently due to causes other than temperature, at least other than temperature alone."

More experiments such as those of Schubel (1974) and Hoss et al. (1973) should be conducted with increased attention on egg development following exposure to excess temperatures, and on the behavioral and physiological responses of larvae to thermal shock --to both increasing and decreasing ΔT's. The experiments by Coutant (1973), and Dean and Coutant (1968) with juvenile salmonids, and by Hoss et al. (1973) with a variety of larval fishes clearly indicate that thermal shock results in a loss of equilibrium and an increase in susceptibility of predators. Even at thermal doses 10 to 20 percent of those required to produce an obvious loss of equilibrium, Coutant (1973) found significant increases in the predation rates on exposed salmonids by larger fish. The physiological response to temperature shock has been studied by Reaves et al. (1968) and Baslow (1967). Reaves et al. (1968) found that acute and moderate thermal shocks of rainbow trout (*Salmo gairdneri*) resulted in decreases in plasma sodium and chloride levels, decreases in tissue water, and an increase in extracellular fluid. Baslow (1967) reported changes in the concentrations of certain amino acids and enzymes in the central nervous systems of fish. The time-temperature exposure histories used by these investigators were not representative of power plants, however.

Entrainment, then, may lead directly to death, or as Coutant (1970b) has pointed out, the sycondary effects such as loss of equilibrium may ultimately affect survival. There may also, of course, be no identifiable response. Coutant's (1971b) application of the principles of "thermal resistance" diagrams (Fry et al., 1946) to entrainment studies provides a useful conceptual model for predicting the effects of entrainment, but many of the data required for application of the model are still lacking. In addition, Hoss et al. (1973) have pointed out that predictions generated by applying Coutant's model to their data did not agree with predictions from their own "shock survival" model.

Considerably more research is required before appropriate maximum permissible ΔT's, or other thermal standards can be set. A number of investigators (e.g., Schubel, 1974; Hoss et al., 1973; Schubel and Pritchard, 1972) have pointed out that thermal standards should be flexible and be based on the most sensitive species to be protected in a particular region. According to this philosophy, standards would vary both spatially and temporally. Maximum permissible ΔT's would probably be lower in summer than in winter, and might be lower in spawning and nursery areas than in other parts of the same water system. The system would be zoned. The ability of a water body to tolerate each "environmental insult" before suffering significant ecological or aesthetic damage varies not only from system to system, but also in different parts of a given system. And, within any segment of a given system it varies temporally. Uniform, invariant standards for the disposal of wastes, whether they are heat, nutrients, or dredged spoil, are environmentally naive. The only justification for their enactment is that it simplifies enforcement. A uniform speed limit of 25 mph is as irrational as one of 100 mph is irresponsible.

There has been a tendency among many biologists to favor low ΔT's and relatively large volumes of cooling water. One might argue for the highest ΔT that can be "safely" used since the higher the ΔT, the smaller the volume of cooling water required. By reducing the volume of cooling water, the mechanical effects of entrainment on the population should be reduced since fewer organisms would be passed through the plant. The effects of biocides on the populations could also be reduced by such a practice. A number of investigators (e.g., Schubel, 1974; Adams, 1968) have indicated that mechanical stresses during entrainment probably have a much greater impact on fish eggs and larvae than does temperature.

The basic question is, of course, not what fractions of the entrained fish eggs and larvae are killed, but rather whether or not the "predation" by the plant will significantly affect the fish

ACKNOWLEDGMENTS

The author's research was supported by the Power Plant Siting Program, Department of Natural Resources, State of Maryland. I thank Mr. Lee Zeni, Director of that program, for his continued cooperation. Preparation of this review paper was supported by the Marine Sciences Research Center of the State University of New York at Stony Brook. Contribution 109 of the Marine Sciences Research Center (MSRC).

LITERATURE CITED

ADAMS, J. R. 1968. Thermal effects and other considerations at steam electric plants. A survey of studies in the marine environment. Pacific Gas and Electric Co., Dept. of Engineering Research, Report No. 6934.4-68.

ADAMS, J. R. 1969. Ecological investigations around some thermal power stations in California tidal waters. Ches. Sci. 10(3-4): 145-154.

ALBRECHT, A. B. 1964. Some observations on factors associated with survival of striped bass eggs and larvae. Calif. Fish and Game 50(2): 101-113.

BASLOW, M. H. 1967. Temperature adaptation and the central nervous system of fish. P. 205-206. In Prosser, C. L. (ed), Molecular mechanism of temperature adaptation. Amer. Assn. Adv. Sci. Publ. No. 84, Washington, D.C.

BRADFORD, A. D., J. G. MILLER and K. BUSS. 1968. Bio-assays on eggs and larva stages of American shad, _Alosa sapidissima_. P. 52-60. In Suitability of the Susquehanna River for Restoration of Shad. Appendix 2. U.S. Dept. of Interior, Washington, D.C.

COMBS, B. D. 1965. Effect of temperature on the development of salmon eggs. Prog. Fish-Cult. _27_(3): 134-137.

COUTANT, C. C. 1968. Thermal pollution--biological effects. A review of the literature of 1967 on wastewater and water pollution control. J. Water. Poll. Cont. Fed. _40_: 1047-1052.

COUTANT, C. C. 1969. Thermal pollution--biological effects. A review of the literature of 1968 on wastewater and water pollution control. J. Water Poll. Cont. Fed. _41_: 1036-1053.

COUTANT, C. C. 1970a. Thermal pollution--biological effects. A review of the literature of 1969 on wastewater and water pollution control. J. Water Poll. Cont. Fed. _42_: 1025-1057.

COUTANT, C. C. 1970b. Biological aspects of thermal pollution 1. Entrainment and discharge canal effects. CRC Critical Reviews in Environmental Control _1_(3): 341-381.

COUTANT, C. C. 1971a. Thermal pollution--biological effects. A review of the literature of 1970 on wastewater and water pollution control. J. Water Poll. Cont. Fed. 43: 1292-1334.

COUTANT, C. C. 1971b. Effects on organisms of entrainment in cooling water: steps toward predictability. Nuclear Safety 12(6): 600-607.

COUTANT, C. C. 1973. Effect of thermal shock on vulnerability of juvenile salmonids to predation. J. Fish. Res. Bd. Canada 30(7): 965-973.

DEAN, J. M. and C. C. COUTANT. 1968. Lethal temperature relations of juvenile Columbia River Chinook salmon. Biological Effects of Thermal Discharges: Annual Progress Report for 1967. Battelle Mem. Inst., Pacific Northwest Lab., Richland, Wash. BNWL-714: 3-8.

EDSALL, T. A. 1970. The effect of temperature on the rate of development and survival of alewife eggs and larvae. Trans. Amer. Fish. Soc. 99: 376-380.

EDSALL, T. A. and T. G. YOCUM. 1972. Review of recent technical information concerning the adverse effects of once-through cooling on Lake Michigan. U. S. Fish and Wildlife Service, Bureau of Sport Fisheries and Wildlife, Great Lakes Fishery Laboratory, Ann Arbor, Mich. 86 p.

EUROPEAN INLAND FISHERIES ADVISORY COMMISSION. 1969. Working paper on water quality criteria for European freshwater fish, water temperature and inland fisheries. Water Research 3: 645-662.

FRANK, MARILYN. 1973. Relative sensitivity of different embryonic stages of carp to thermal shock. In Gibbons, J. W. and R. R. Sharitz (eds), Thermal Ecology AEC Symposium Series. Conf.-730505, in press.

FRY, F. E. J., J. S. HART, and K. F. WALKER. 1946. Lethal temperature relations for a sample of young speckled trout (*Salvelinus fontinalis*). Univ. of Toronto Studies, Biological Series, Pub. Ont. Fish. Res. Lab. 66: 5-35.

GUNTER, G. and J. W. WARD. 1961. Some fishes that survive extreme injuries, and some aspects of tenacity of life. Copeia (4): 454-462.

HATHAWAY, E. S. 1927. The relations of temperature to the quantity of food consumed by fish. Ecology 8: 428-434.

HAYES, F. R. and F. H. ARMSTRONG. 1942. Physical changes in the constituent parts of developing salmon eggs. Can. J. Res. D. 20: 99-114.

HOSS, D. E., W. F. HETTLER JR., and L. C. COSTON. 1973. Effects of thermal shock on larval estuarine fish--Ecological implications with respect to entrainment in power plant cooling systems. In the Proceedings of the Symposium on the Early Life History of Fish. Oban, Scotland.

HUNTSMAN, A. G. and M. I. SPARKS. 1924. Limiting factors for marine animals. 3. Relative resistance to high temperature. Contr. Can. Biol. Fish. N.S. 2: 102-113.

KELLEY, J. W. 1968. Effects of incubation temperature on survival of largemouth bass eggs. Prog. in Fish Culture 30: 159-163.

KENNEDY, U.S. and J. A. MIHURSKY. 1967. Bibliography of the effects of temperature in the aquatic environment. Univ. of Maryland Natural Resources Institute, Contrib. No. 326, Solomons, Md. 89 p.

KERR, J. E. 1953. Studies on fish preservation at the Contra Costa Steam Plant of the Pacific Gas and Electric Company. Fish Bull. No. 92, Calif. Dept. Fish and Game.

LEE, W. S. 1970. Considerations in translating environmental concern into power plant design and operation. Paper presented at annual meeting of Atomic Industrial Forum, Washington, D.C., June 28-30, 1970.

MARCY, B. C. JR. 1969. Resident fish population dynamics and early life history studies of the American shad in the lower Connecticut River, The Connecticut River Investigation, 9th Semiannual Progress Rept., p. 13-32, October 1969.

MARCY, B. C. JR. 1971. Survival of young fish in the discharge canal of a nuclear power plant. J. Fish. Res. Bd. of Canada 28(7): 1057-1060.

MARCY, B. C. JR. 1973. Vulnerability and survival of young Connecticut River fish entrained at a nuclear power plant. J. Fish. Res. Bd. Canada 30(8): 1195-1203.

MARKOWSKI, S. 1959. The cooling water of power stations, a new factor in the environment of marine and fresh-water invertebrates. J. Anim. Ecol. 28(2): 243-258.

PRICE, J. W. 1940. Time-temperature relations in the incubation of the whitefish, Coregonua clupeaformis (Mitchill). J. Gen. Physiology 23: 449-468.

PRITCHARD, D. W. and H. H. CARTER. 1972. Design and citing criteria for once-through cooling systems based on a first-order thermal plume model. Tech. Report 75, Reference 72-6 of the Chesapeake Bay Institute, The Johns Hopkins Univ., Baltimore, Md. 51 p.

RANEY, E. C., B. W. MENZEL and E. C. WELLER. 1973. Heated effluents and effects on aquatic life with emphasis on fishes, a bibliography. Ichthyological Associates Bull. No. 9, U.S. Atomic Energy Comm., Office of Info. Services, Tech. Info. Center. Available from Natl. Tech. Info. Serv., Springfield, Va. 22151.

REAVES, R. S., A. H. HOUSTON, and J. A. MADDEN. 1968. Environmental temperature and the body fluid system of the freshwater Telostii. Ionic Regulation in Rainbow Trout, Salmo gairdneri, Following Abrupt Thermal Shock. Comp. Biochem, Physiol. 25(3): 849-860.

SCHIEMER, E. W. and J. R. SCHUBEL. 1974. Constant temperature water supply system. Special Report 39, Reference 74-9 of the Chesapeake Bay Institute, The Johns Hopkins University, Baltimore, Md. 10 p.

SCHUBEL, J. R. 1974. Effects of exposure to time-excess temperature histories typically experienced at power plants on the hatching success of fish eggs. Estuarine and Coastal Mar. Sci. 2: 105-116.

SCHUBEL, J. R. and A. H. AULD. 1972a. Thermal effects of a model power plant on the hatching success of American shad, Alosa sapidissima, eggs. Proceedings 26th Annual Conf. Southeastern Assoc. Game & Fish Comm., Knoxville, Tennessee, 22-25 October 1972, in press.

SCHUBEL, J. R. and A. H. AULD. 1972b. Thermal effects of a model power plant on the hatching success of alewife (Alosa pseudoharengus) eggs. Special Report 28, Ref. 72-14 of the Chesapeake Bay Institute of The Johns Hopkins Univ., Baltimore, Md. 13 p.

SCHUBEL, J. R. and A. H. AULD. 1973a. Effects of exposure to time-excess temperature histories typically experienced at Power plants on the hatching success of blueback herring

(Alosa aestivalis) eggs. Special Report 31, Ref. 73-4 of the Chesapeake Bay Institute of The Johns Hopkins Univ., Baltimore, Md. 17 p.

SCHUBEL, J. R. and A. H. AULD. 1973b. Effects of exposure to time-excess temperature histories typically experienced at power plants on the hatching success of blueback herring and striped bass eggs. In J. W. Gibbons and R. R. Sharitz (eds.), Thermal Ecology AEC Symposium Series. Conf.-730505, in press.

SCHUBEL, J. R. and D. W. PRITCHARD. 1972. The estuarine environment Part 2. The J. of Geol. Ed. 20(4): 179-188.

SHANNON, E. H. 1970. Effect of temperature changes upon developing striped bass eggs and fry. Proceedings of the 23rd Ann. Conf. of Southeastern Assoc. Game & Fish Comm., 19-22 October 1969, p. 265-274.

SWARUP, H. 1958. Abnormal development in the temperature-treated eggs of Gasterosteus aculeatus (L.). I. Cleavage Abnormalities. J. Zool. Soc. India 10: 108-113.

SWARUP, H. 1959a. Abnormal development in the temperature-treated eggs of Gasterosteus aculeatus (L.). II. Gastrulation Abnormalities. J. Zool. Soc. India 11: 1-6.

SWARUP, H. 1959b. Abnormal development in the temperature-treated eggs of Gasterosteus aculeatus (L.). III. Twinning. J. Zool. Soc. India 11: 7-10.

U. S. ATOMIC ENERGY COMMISSION. 1969. Thermal effects studies by nuclear power plant licenses and applicants. U. S. Government Printing Office, Washington, D.C.

VOLODIN, V. M. 1960. Effect of temperature on the embryonic development of the pike, the blue bream (Abramis ballerus L.) and the white bream (Blicca bjoerkna L.). Ipst. Cat. No. 941 (Transl. from Russian).

WATER RESOURCES COUNCIL. 1968. The nation's water resources. U. S. Government Printing Office, Washington, D.C.

ENVIRONMENTAL MONITORING THROUGH THE USE OF EXPOSURE PANELS

Ribert E. Hillman
Battelle
William F. Clapp Laboratories, Inc.
Duxbury, Massachusetts

ABSTRACT

Exposure panels made of white pine with a transite facing are useful for assessing population and community changes in both polluted and unpolluted waters. Their uniform size and shape and the ease with which they can be handled makes them ideal for sampling a wide variety of sessile plants and animals, and those motile forms usually associated with sessile communities.

Data from an ecological study at Millstone Point are used as examples of the kinds of information provided through the use of exposure panels as environmental monitors. Species diversity indexes for both algal and annelid populations as they occurred on panels from 1968 through 1973 were calculated. Regression analyses over time showed increases in diversity indexes since the program began, indicating possible changes in water quality.

For the period of October 1971 through September 1972, the number of phyla found on panels exposed at Millstone Point, New Haven Harbor, and Stamford Harbor were not significantly different, but far more species were found on the Millstone Point panels.

INTRODUCTION

The inclusion of a discussion of the use of exposure panels, which traditionally have been used to monitor cycles of fouling organisms and borers relative to biodegradation of ships, wharves, pilings, and other structures placed in the marine environment, in a symposium dealing with fisheries and energy production, may be questionable, although modified panel-type studies have been used to assess commercial shellfish populations, or at least the potential for harvest, based upon the number of larvae which "set" on these panels (e.g., Miyazaki, 1938; Butler, 1954; Shaw, 1967).

Such panels have been used by some electric utilities to determine the presence and abundance of organisms which might foul intake pipes and tunnels, thereby reducing the efficiency of the cooling system, but they were not generally used to provide ecological information on the effects of energy production on fisheries. Recently, however, a number of investigators have recognized that the degree of complexity of the communities which grow on these panels has been indicative of the general quality of the environment in which the panels have been exposed, and that these panels could be used to help explain some of the changes in the fish populations inhabiting the same waters. It is recognized, of course, that they have certain limitations and would not answer all of the questions, but they would go a long way toward defining possible changes in the environment.

The assemblage, or community, of fouling organisms which is able to develop over a period of time depends on several factors, chief among them being seasonality, or the availability of life cycle stages of the fouling forms usually present in the area at the time the clean surface is immersed in the sea. Of interest here are water quality factors, e.g., temperature, salinity, pH, nutrients, trace metals, dissolved oxygen, and turbidity, which limit or control the availability of the fouling organisms. These factors also dictate, to a large extent, which fish species can survive in the area. For example, if a highly polluted area can support the growth of only a few fouling species, in all likelihood, the number of fish species found in the area will also be limited. It is this principle on which the use of exposure panels as environmental monitors is based. By assessing the assemblage of fouling organisms and borers over a period of time, one can define existing conditions in terms of species abundance and diversity and community stability, and then use possible changes in these same parameters to detect any changes in the environmental factors which may have caused the changes.

The animals and plants which make up a fouling community are primarily the sessile, or attached, species which occur naturally in the coastal waters. There are, however, many free-living organisms such as polychaetes, amphipods, and gastropods which are integral parts of the fouling community. It is this tremendous diversity, and the ease with which it can be followed, that makes the use of exposure panels a valuable tool for monitoring the effects of power plants on the environment, and, thereby, on fish populations.

In a publication by the Woods Hole Oceanographic Institution (1952), prepared for the U.S. Navy, it was stated that nearly 2,000 species (615 plants and 1,361 animals) have been reported from fouling communities. This list included almost all phyla, with certain exceptions like the pelagic Ctenophora and Chaetognatha, the phylum Nematoda, which almost certainly is present but at that time had not been identified, and the phylum Phoronidea, which is sessile and, like the Nematodes, may ultimately be recorded.

Of the nearly 2,000 species listed by the Woods Hole Oceanographic Institution, 464 appeared on artificial test surfaces. The use of blocks or panels to investigate fouling communities apparently dates back at least to Dahl (1893), who used them in investigations of the fauna of the lower Elbe River.

The literature on the development of fouling and borer communities, on both natural and artificial substrates, is quite extensive, and will not be detailed here. For the interested reader, excellent summaries and bibliographies are provided by the Woods Hole Oceanographic Institution (1952), Clapp and Kenk (1963), Frame (1968), and Schoener (1974), who discusses an interesting thesis that the artificial substrates represent mini-islands undergoing colonization.

A number of investigators have recently used the exposure panel, or test board technique to monitor the effects of power plant construction and operation on the environment (see e.g., Markowski, 1960; Cory, 1967; Frame, 1968; Nauman and Cory, 1969; Fairbanks et al., 1971; Hillman et al., 1973). Although the specific techniques varied, the principle in each case was basically the same. Panels of various materials, e.g., wood, asbestos, and cement, were immersed in sea water at locations which were in or around the area of the condenser cooling water intake structures and/or affected by the thermal effluent, and in control locations not affected by the power plant. The panels were immersed for various periods of time, and then the flora and fauna growing on them were identified and quantified in some manner to provide information on the abundance and composition of the fouling communities at the various locations.

The William F. Clapp Laboratories of Battelle's Columbus Laboratories have been using the exposure panel technique for several years as part of more comprehensive programs for monitoring the effects of power plants on Long Island Sound. Panel racks have been immersed at locations in Stamford Harbor, New Haven Harbor, and around Millstone Point since the late 1960's. The most extensive study of the three sites is currently underway around Millstone Point at the eastern end of Long Island Sound, and has been in progress since June 1968. To illustrate the effectiveness of the use of exposure panels for monitoring purposes, I would like to discuss in some detail the results, to date, of the Millstone Point exposure panel program, and make some comparisons of the Millstone Point data with data from Stamford Harbor at the western end of the Sound, and with data from New Haven Harbor, which is about midway between the other two sites. Data comparisons will be from comparable time periods. The techniques of exposure were essentially the same for all sites.

THE MILLSTONE POINT STUDY

Panels were set out at White Point, Fox Island-North, and Millstone Harbor (Fig. 1) in June 1968. An additional rack was installed at Black Point in October 1968, but was moved to Giant's Neck in January 1969 because of difficulties with the installation. A fifth rack was installed at the intake in April 1969. In July 1973, an exposure panel unit was installed in the effluent quarry, suspended from a platform along the eastern bank of the quarry.

Each unit was installed immediately below low-tide level, thus subjecting it to minimum variation of depth-related parameters and maximum impact of the warm water effluent. The unit consisted of a series of untreated southern white pine wood/transite panels (Fig. 2). Each month, two panels were removed--one that had been exposed for one month (the control) and one that had been exposed for a 12-month period. The short-term, or control, panels enabled the number of species which attached each month to be determined. The long-term panels allowed measurement of growth over extended periods of time and enabled determination of annual and seasonal variations due to fluctuations in water temperature and/or other environmental factors. Each panel, upon arrival at the laboratory, was held in running seawater for not more than three days, until examined. A subjective determination of the percent of the panel covered by a given species was made (±5 percent). Where possible, a detailed enumeration of all macroscopic biota was also made. The growth rates of the fouling and boring organisms were noted where possible.

Throughout most of the reporting period, it was extremely difficult to keep a rack at the intake because of wave action and surge. It was decided to anchor the rack approximately 100 feet in front of the intake screens. A 50-pound mushroom anchor with 40 feet of chain was attached to the rack, and the rack was floated just beneath the surface with 2 heavy-duty floatation buoys. This arrangement has, thus far, been quite successful.

At the time of the preparation of this manuscript, over 170 species, or taxonomic groups, representing 12 phyla, have appeared on the panels in the Millstone Point area. Most of these occurred rarely (i.e., only once) or uncommonly (i.e., on no more than 5 percent of the panels). Sixty-five species, including 17 species of algae and 48 species of animals, from sponges through chordates, appeared prior to the operation of the plant.

The beginning of plant operations in December 1970 marked the end of the preoperational phase of the study, and was a convenient

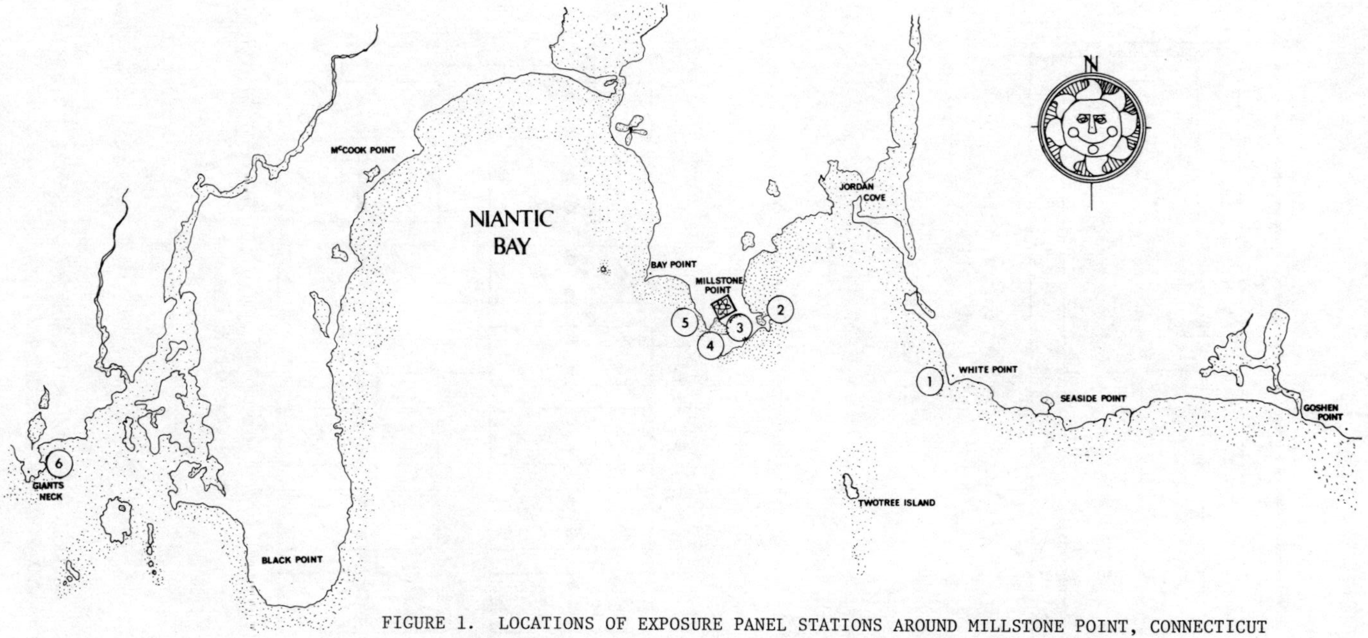

FIGURE 1. LOCATIONS OF EXPOSURE PANEL STATIONS AROUND MILLSTONE POINT, CONNECTICUT

1. White Point 2. Fox Island 3. Effluent 4. Millstone Harbor
5. Intake 6. Giant's Neck

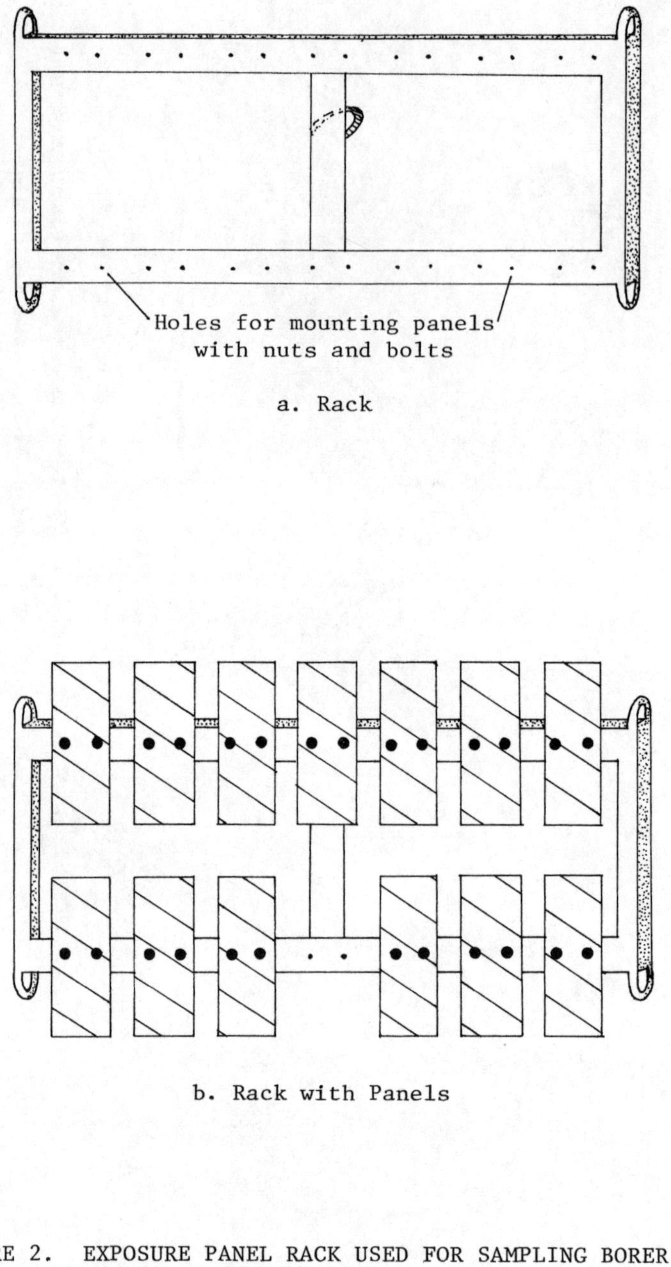

FIGURE 2. EXPOSURE PANEL RACK USED FOR SAMPLING BORER AND FOULING ORGANISMS

point to analyze any trends in occurrence of fouling organisms as collected on the exposure panels from June 1968 to that point. Prior to plant operations, the most prevalent algal forms were the several species of Enteromorpha, Ulva lactuca, Codium fragile, and the red algae, Ceramium rubrum. The ectoproct, Cryptosula pallasiana and the barnacles, Balanus crenatus and B. eburneus, were the most abundant fouling animals.

Because of their persistent presence on the panels, or their obvious seasonal patterns, 24 species were chosen for more concentrated study. A Chi-square contingency analysis was run on the data to determine whether there were any significant relationships between the presence or abundance of the species and the month or year collected, or the site at which the organism occurred. The results of that pre-operational analysis are summarized in Table 1. The hypothesis was made that the organisms' occurrences were not dependent upon each of these factors. The hypothesis was rejected if the chance of the organisms' being dependent on any one of the factors was 90 percent or better (significance in the Table equalling 10 percent or less).

A similar analysis was made for the same species after the plant had been in operation for about two years, and a summary of the results of that analysis is shown in Table 2.

Several trends, or shifts in species composition, as noted on the exposure panels, have taken place since the program started in 1968. For example, the red alga, Ceramium rubrum, pre-operationally the most consistently present red alga on the panels, has not appeared since 1971. Likewise, the various species of the green algal genus Enteromorpha have declined in abundance since 1969, to the point where their occurrence on the panels is uncommon. Sertularia pumila, a hydroid, was originally found to be most abundant during the spring and summer. Though still present at about the same level year round, it began to decline in abundance, and by 1972 had disappeared from all exposure panel sites.

While certain species declined, others increased in abundance on the panels. The mussel, Mytilus edulis, for example, appeared only sporadically on the panels during the pre operational phase of the study, but now appear frequently at most sites. This corresponds with general observations of increases in Mytilus along the rocky shores of the Millstone Point area, and is of interest in this symposium because the mussel forms one of the principal foods of the tautog, Tautoga onitis, a common sport fish in the area.

For the most part, the trends in abundance of the various species appear reflective of natural cycles rather than of the effect

Table 1. List of species found during pre-operational phase to be dependent upon the month, the site, or the year in which they occurred on panels in the Millstone Point, Connecticut, area.

(L) Indicates species from long-term panels
(S) Indicates species from short-term panels

Month	Site	Year
Enteromorpha compressa (S)	*Enteromorpha linza* (L)	*Ulva lactuca* (L)
Enteromorpha intestinalis (L)	*Ulva lactuca* (L)	*Certularia pumila* (L)
Enteromorpha intestinalis (S)	*Codium dichotomum* (L)	*Cryptosula pallasiana* (L)
Enteromorpha linza (L)	*Ceramium rubrum* (L)	*Bugula turrita* (L)
Enteromorpha spp. (S)	*Bugula turrita* (L)	*Bugula turrita* (S)
Ceramium rubrum (S)	*Mytilus edulis* (L)	*Mytilus edulis* (L)
Halichondria bowerbanki (L)	*Mytilus edulis* (S)	*Mytilus edulis* (S)
Sertularia pumila (L)	*Teredo navalis* (L)	*Teredo navalis* (L)
Sertularia pumila (S)	*Balanus crenatus* (L)	*Lepidonotus squamatus* (L)
Cryptosula pallasiana (S)	*Balanus crenatus* (S)	*Balanus crenatus* (L)
Bugula simplex (L)	*Balanus improvisus* (L)	*Balanus eburneus* (L)
Bugula simplex (S)	*Limnoria* tunnels (L)	*Limnoria* tunnels (L)
Mytilus edulis (L)	*Limnoria* tunnels (S)	*Corophium cylindricum* (L)
Balanus eburneus (S)		
Balanus spp. (S)		
Limnoria tunnels (S)		
Botryllus schlosseri (L)		
Botryllus schlosseri (S)		

Table 2. List of species found during the operational phase to be dependent upon the month, the site, or the year in which they occurred on panels in the Millstone Point, Connecticut, area.

(L) Indicates species from long-term panels
(S) Indicates species from short-term panels

Month	Site	Year
Enteromorpha spp. (L)	*Enteromorpha intestinalis* (L)	*Enteromorpha compressa* (L)
Enteromorpha spp. (S)	*Ulva lactuca* (L)	*Enteromorpha intestinalis* (L)
Halichondria bowerbanki (L)	*Codium dichotomum* (L)	*Enteromorpha* spp. (L)
Cryptosula pallasiana (S)	*Halichondria bowerbanki* (L)	*Ceramium rubrum* (L)
Bugula simplex (L)	*Sertularia pumila* (L)	*Sertularia pumila* (L)
Bugula simplex (S)	*Cryptosula pallasiana* (L)	*Sertularia pumila* (S)
Bugula turrita (L)	*Bugula turrita* (L)	*Teredo navalis* (L)
Teredo navalis (L)	*Mytilus edulis* (L)	*Balanus crenatus* (L)
Balanus eburneus (S)	*Mytilus edulis* (S)	*Balanus eburneus* (L)
Balanus spp. (L)	*Teredo navalis* (L)	*Balanus eburneus* (S)
Balanus spp. (S)	*Lepidonotus squamatus* (L)	*Limnoria* tunnels (L)
Limnoria tunnels (L)	*Balanus crenatus* (L)	
Limnoria tunnels (S)	*Balanus eburneus* (L)	
Botryllus schlosseri (S)	*Balanus improvisus* (L)	
	Limnoria tunnels (L)	
	Limnoria tunnels (S)	
	Chelura terebrans (L)	
	Corophium cylindricum (L)	
	Botryllus schlosseri (L)	

of the power plant, since they have occurred throughout the area, including the control site where no influence of the thermal discharge is felt.

Stress on natural populations due to poor water quality is often reflected in decreases in species diversity. Wilhm (1967), for example, showed that decreases in diversity indexes may be correlated with pollution-caused population changes. Nauman and Cory (1969), on the other hand, calculated epifauna diversity indexes from exposure panels suspended in the thermal discharge of a power station on the Patuxent River in Maryland and showed that effluent populations of macrofauna had, on the average, higher diversities than the intake populations.

Since the commonly used diversity indexes require the number of individuals of each species, it is not possible to calculate a single index for all populations on the panels, because of the difficulty in enumerating such organisms as algae, bryozoans, hydroids, and other similar forms. It is possible to calculate diversity indexes for populations of macroorganisms like the annelids, however, and to use them in comparisons between sites and over time. This point is illustrated below.

Table 3 shows the number of individuals of each of the annelid species collected at each station each year since the program began. Diversity indexes were calculated using the Shannon-Wiener $\hat{H}$ formula (see e.g., Sanders, 1968),

$$H = - \sum_{i=1}^{S} \frac{Ni}{N} \log_2 \frac{Ni}{N},$$ for each site on a yearly basis. These were determined from the total numbers of individuals of each species found during the year.

The data show an increase in both the number of species and the numbers of individuals of each species in the Millstone Point area since the program began. To determine whether there was a true increase in diversity since the start of the program, correlation coefficients were calculated, to determine whether diversity increased over time. In all cases, there was a significant positive correlation between the increase in annelid diversity index and the increase in time, i.e., from year to year, of the study (Table 4).

Since the algae represented the largest single group of organisms, in terms of number of species (see Table 5), similar diversity indexes were calculated for that group. Because the abundance of algae could not be quantified in terms of individuals, but rather

Table 3. Abundance of annelid species as found on exposure panels located in the Millstone Point, Connecticut, area.

[1]Panels exposed from June through December
[2]Data through July
[3]No panel for most of 1968

	1968[1]	1969	1970	1971	1972	1973	1974[2]
				White Point			
Eulalia sp.							5
Eumida sp.						1	
Glyceridae							1
Lepidonotus squamatus			25	24	10	16	9
Nephtys sp.			1				
Nereis succinea						4	1
Nereis virens		16	27	10	4	4	
Notomastus latericius							1
Phyllodocidae					1	6	8
Podarke obscura							1
Sabellidae					9	11	27
Terebellidae			1				3
				Fox Island			
Ampharetidae				1			
Capitellidae							1
Cirratulidae						1	
Eulalia sp.						3	5
Eumida sp.						3	6
Lepidonotus squamatus			20	26	10	55	54
Nereis succinea				1		1	10
Nereis virens	3	11	26	7	6	1	6
Phyllodicidae						7	4

Table 3, continued.

	1968[1]	1969	1970	1971	1972	1973	1974[2]
Fox Island							
Platynereis megalops			1				
Sabellidae					2	11	44
Terebellidae					1	1	7
Millstone Harbor							
Euchone rubrocincta						1	
Eulalia sp.							2
Lepidonotus squamatus			11	11	8	32	33
Nereis succinea					3	10	34
Nereis virens	2	7	10	7			
Polynoidae						1	
Sabellidae					1	11	20
Terebellidae					1		3
Intake							
Eulalia sp.							2
Lepidonotus squamatus				4		5	1
Nereis succinea						31	7
Nereis virens			5	12			
Phyllodocidae							1
Sabellidae						2	2
Terebellidae							1
Giant's Neck[3]							
Eulalia viridis					1	2	1
Eumida						2	1
Lepidonotus squamatus			11		10	33	10
Nereis pelagica						1	
Nereis virens		7	7	6	1	1	
Phyllodocidae						1	
Sabellidae						4	9
Terebellidae							200

Table 4. Determination of correlation coefficients for annelid species diversity indexes as calculated from exposure panels in the Millstone Point, Connecticut, area against year for which the index was determined.

	Year	White Point	Fox Island	Millstone Harbor	Giant's Neck
	1969	0	0	0	0
	1970	1.228	1.115	0.999	0.964
	1971	0.874	1.076	0.964	0
	1972	1.679	1.578	1.489	0.817
	1973	2.213	1.735	1.572	1.403
Correlation Coefficient		0.9212	0.9163	0.9179	0.7583
Significance Level		>.001	>.001	>.001	.02>p>.01

Table 5. List of algal species occurring on exposure panels in the Millstone Point, Connecticut, area from June 1968 through June 1974.

	CHLOROPHYTA
Bryopsis plumosa	*Enteromorpha intestinalis*
Chaetomorpha sp.	*Enteromorpha linza*
Cladophora sp.	*Enteromorpha prolifera*
Codium fragile	*Enteromorpha* spp.
Enteromorpha clathrata	*Ulva lactuca*
Enteromorpha compressa	
	RHODOPHYTA
Achrochaetium sp.	*Gracilaria foliifera*
Agardhiella tenera	*Gracilaria* spp.
Antithamnion sp.	*Grinellia americana*
Callithamnion byssoideum	*Herposiphonia tenella*
Callithamnion spp.	*Lomentaria baileyana*
Ceramium diaphanum	*Lomentaria* spp.
Ceramium rubrum	*Polysiphonia nigrescens*
Ceramium sp.	*Polysiphonia* spp.
Champia parvula	*Prophyra umbilicus*
Chondria sp.	*Porphyra* spp.
Cystocloneum purpureum	*Rhodomela subfusca*
Daysa pedicellata	*Rhodymenia palmata*
Gracilaria confervoides	*Rhodymenia* spp.
	PHAEOPHYTA
Desmarestia viridis	*Petalonia fascia*
Elachistea sp.	*Punctaria latifolia*
Ectocarpus sp.	*Pylaiella litoralis*
Laminaria agardhii	*Fucus evanescens*
Laminaria spp.	*Fucus* spp.

in terms of the percent of the exposure panels covered by a particular species, percent coverage was used in the Shannon-Wiener formula in place of the term for number of individuals (cf., Wilhm, 1968). The calculated diversity indexes are shown in Table 6. When regression analyses were carried out, plotting the increase in diversity over time, there was a significant positive correlation between the increase in diversity index and the increase in time.

There are severe criticisms of the use of species diversity indexes to quantify the interrelationships of species within a community (Hurlburt, 1971), and it is not the intent of this author to enter into the controversy. However, we believe that under certain conditions, such as in investigating population differences in polluted versus non-polluted areas, diversity indexes can be useful tools. The purpose of this paper is to discuss the use of an exposure panel program for monitoring the environment in which important fisheries are found. From empirical observations in this case a species diversity index seems workable.

If a species diversity index is reflective of water quality conditions, as has been stated by a number of investigators (e.g., Wilhm and Dorris, 1966; Bechtel and Copeland, 1970; McErlean et al., 1973), it would appear that water quality in the Millstone Point area has improved, or at least changed enough to allow more species to be present, over the past several years.

There is some evidence of long-range warming of the North Atlantic Ocean (Ruddiman et al., 1970) and data compiled at our own laboratories in Duxbury, Massachusetts, suggest that the mean summer monthly and annual temperatures have increased in that area over the past three years. It is also possible, though somewhat optimistic, to state that the increases in diversity indexes point to the efficiency of water pollution control measures in the State of Connecticut.

EXPOSURE PANEL PROGRAMS AT OTHER SITES IN LONG ISLAND SOUND

Exposure panels racks identical to those described above for the Millstone Point study, were placed at locations in Stamford Harbor (Fig. 3) from October 1971 to September 1972, and in New Haven Harbor (Fig. 4) from August 1971 to the present time. For comparative purposes, it is interesting to look at the assemblages of organisms on the three sets of panels for the same time period. Figure 5 is a histogram depicting the phyla and the numbers of

Table 6. Determination of correlation coefficients for algal species diversity indexes as calculated from exposure panels in the Millstone Point, Connecticut, area against year for which the index was determined.

*No rack for most of year

	Year	White Point	Fox Island	Millstone Harbor	Intake	Giant's Neck
	1969	1.956	2.374	1.860	2.095	2.236
	1970	2.424	1.971	2.836	1.944	1.976
	1971	3.260	2.525	2.736	2.000	2.015
	1972	2.762	3.271	2.714	*	2.659
	1973	3.267	2.792	2.926	1.116	2.754
Correlation Coefficient		0.959	0.649	0.739	-0.911	0.753
Significance Probability		>.001	.05>p>.01	.02>p>.01	>.001	.02>p>.01

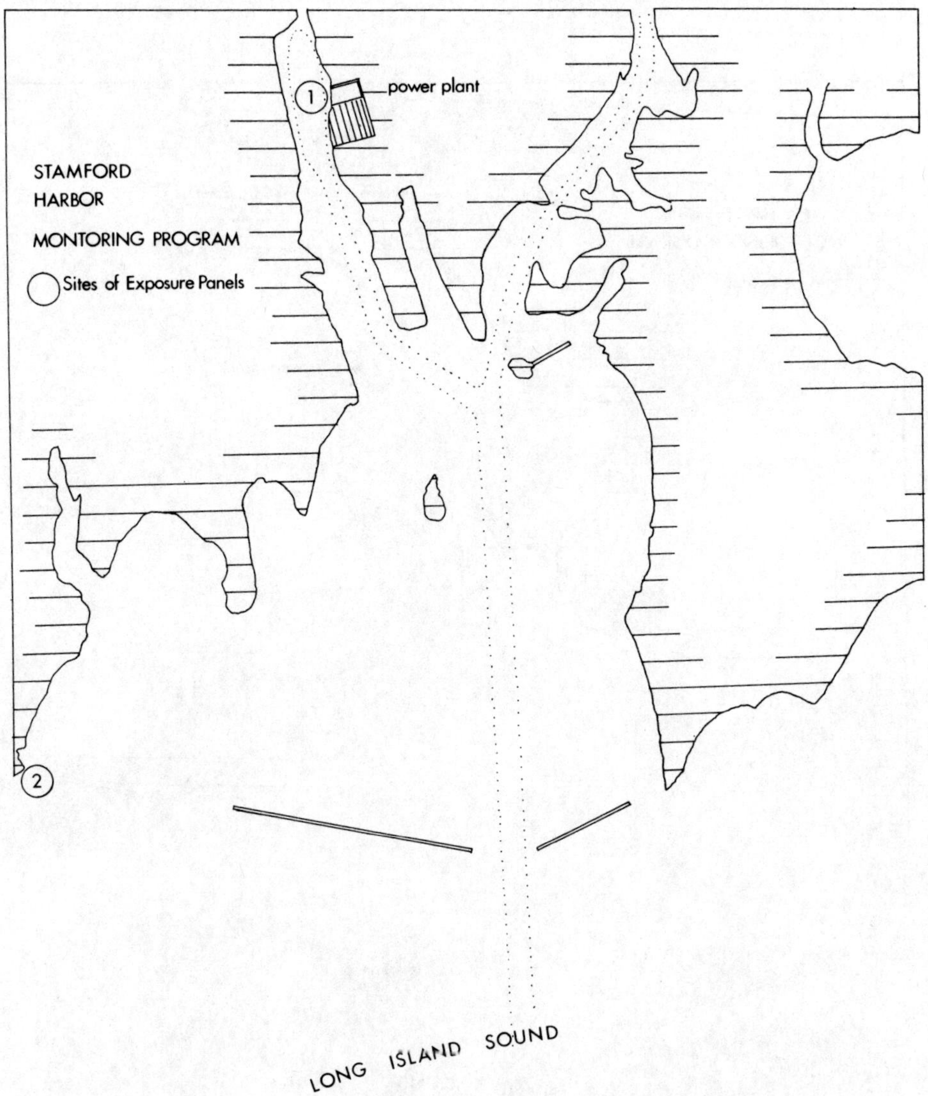

FIGURE 3. LOCATION OF EXPOSURE PANEL STATIONS IN STAMFORD HARBOR

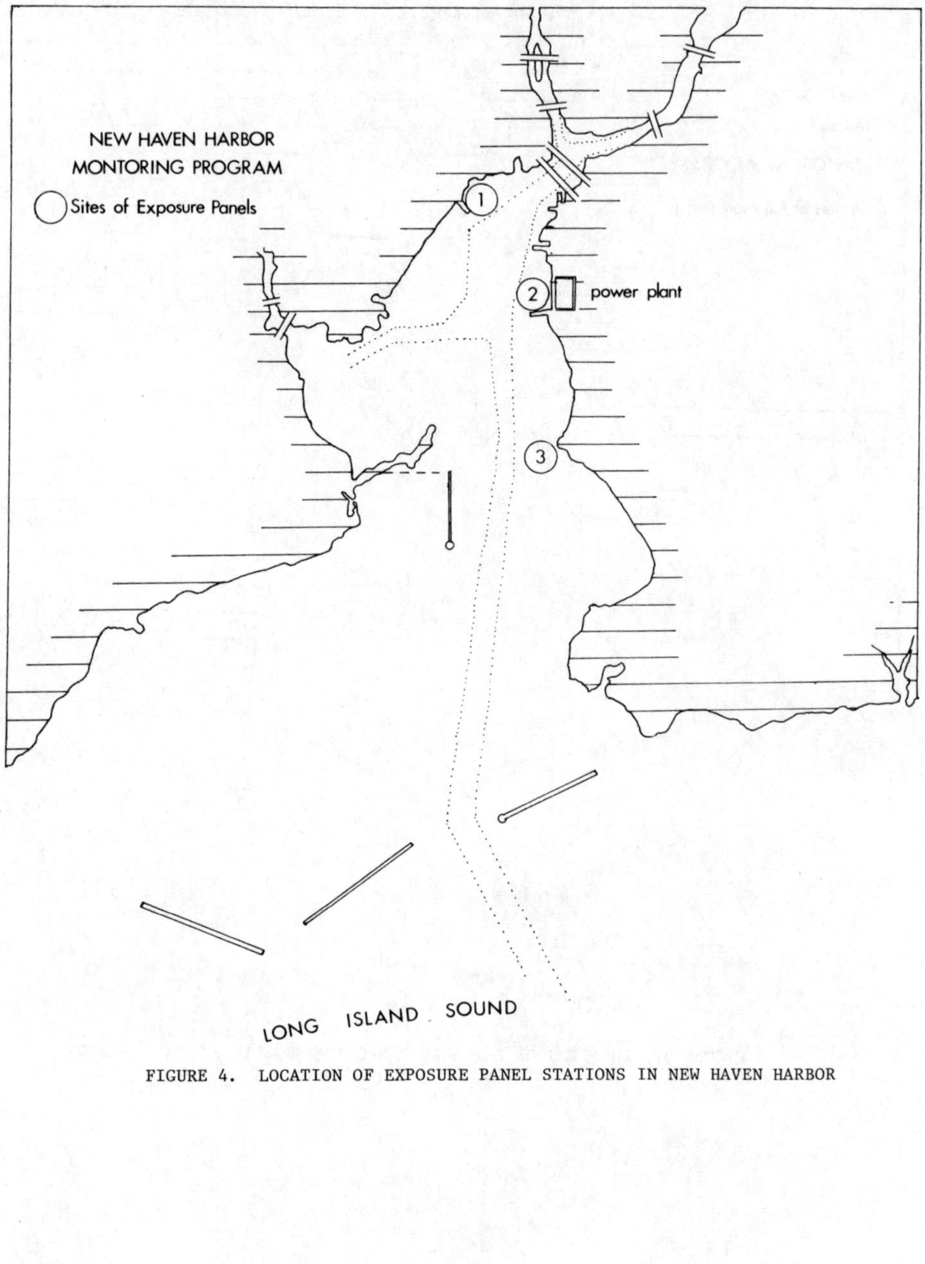

FIGURE 4. LOCATION OF EXPOSURE PANEL STATIONS IN NEW HAVEN HARBOR

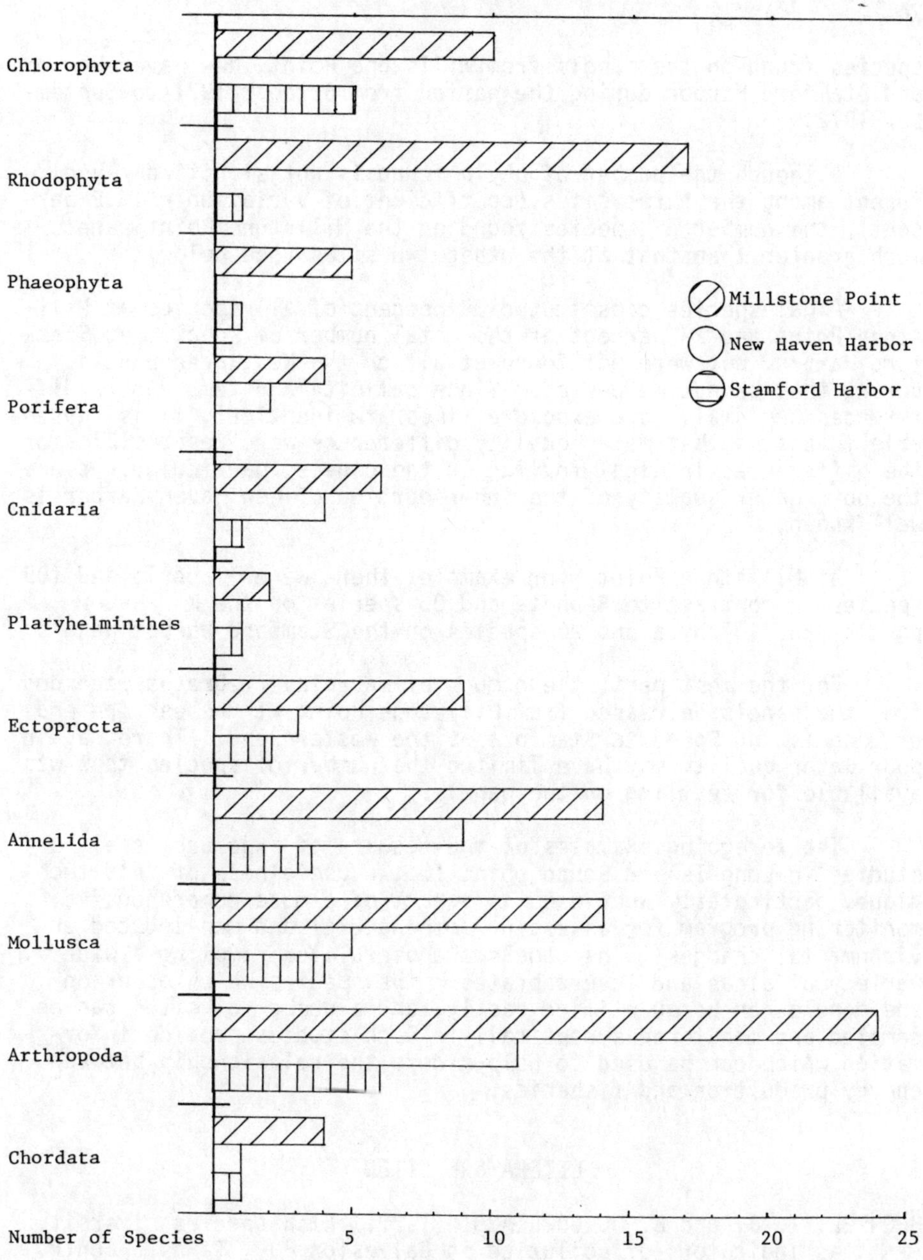

FIGURE 5. A COMPARISON OF THE NUMBER OF SPECIES PER PHYLUM FOUND ON PANELS EXPOSED AT MILLSTONE POINT, NEW HAVEN HARBOR, AND STAMFORD HARBOR FROM OCTOBER, 1971 TO SEPTEMBER, 1972

species found on the panels from Millstone Point, New Haven Harbor, and Stamford Harbor during the period from October 1971 to September 1972.

Although the number of phyla found is not significantly different among the three sites (coefficient of variation = 15.8 percent), the number of species found on the Millstone Point panels is much greater than that at the other two sites (see below).

Algal species constituted 29 percent of all species at Millstone Point and 27 percent of the total number of species at Stamford Harbor, but were not found at all on the New Haven panels during that same time period. Since salinity and temperature differences are small, and exposure times are identical, it is reasonable to assume that water quality differences were responsible for the differences in algal fouling on the panels, particularly since the poor water quality of the inner portion of New Haven Harbor is well known.

At Millstone Point, for example, there were 11 phyla and 109 species in contrast to 8 phyla and 36 species on the New Haven panels, and 10 phyla and 26 species on the Stamford Harbor panels.

For the most part, the number of macroinvertebrates recorded from the panels decreased from Millstone Point at the eastern end of Long Island Sound to Stamford at the western end. There, again, poor water quality may have limited the number of species that was available for settling on the panels.

The foregoing examples of the results of exposure panel studies in Long Island Sound point to the usefulness of this technique, particularly when used as a part of a more comprehensive monitoring program for assessing both natural and man-induced environmental changes. The panels are useful for sampling a wide variety of algae and invertebrates. The species which occur on the panels can be quantified easily, and a number of sites can be sampled and monitored synoptically. Such studies provide information which can be used to help assess the relationship between energy production and fisheries.

LITERATURE CITED

BECHTEL, T. J. and B. J. COPELAND. 1970. Fish species diversity as indicators of pollution in Galveston Bay, Texas. Contr. in Mar. Sci. 15: 103-132.

BUTLER, P. A. 1954. Selective setting of oyster larvae on artificial cultch. Proc. Nat. Shellfish. Assoc. 45: 95-105.

CLAPP, W. F. and R. KENK. 1963. Marine Borers. An Annotated Bibliography. Office of Naval Research, Washington, D.C. 1136 pp.

CORY, R. L. 1967. Epifauna of the Patuxent River Estuary, Maryland, for 1963 and 1964. Ches. Sci. 8: 71-89.

DAHL, F. 1893. Untersuchungen über die Thierwelt det Unterelbe: Jahresb. Commiwis. unters. deuts. Meerekiel. 6: 151-185.

FAIRBANKS, R. B., W. S. COLLINGS, and W. T. SIDES. 1971. An Assessment of the Effects of Electrical Power Generation on Marine Resources in the Cape Cod Canal. Mass. Dept. Nat. Res., Div. Mar. Fish. 48 pp.

FRAME, D. W. 1968. Epibiota attached to wooden panels in the Cape Cod Canal, Sandwich, Massachusetts. M. S. Thesis, Univ. of Mass. 146 pp.

HILLMAN, R. E., N. W. DAVIS, and R. W. COTE. 1973. A Monitoring Program on the Ecology of the Marine Environment of the Millstone Point, Connecticut Area, with Special Attention to Key Indicator Organisms - Pre-Operational Phase. Summary Report to Northeast Utilities Service Company. Battelle, Columbus Laboratories, William F. Clapp Laboratories, Duxbury, Mass.

HURLBURT, S. H. 1971. The nonconcept of species diversity: a critique and alternate parameters. Ecology 52: 577-586.

MARKOWSKI, S. 1960. Observations on the response of some benthonic organisms to power stations cooling water. J. Animal Ecol. 29: 349-357.

McERLEAN, A. G., S. G. O'CONNOR, J. A. MIHURSKY, and C. I. GIBSON. 1973. Abundance diversity, and seasonal patterns of estuarine fish populations. Estuarine and Coastal Mar. Sci. 1: 19-36.

MIYAZAKI, I. 1938. On fouling organisms in the oyster farm. Bull. Jap. Soc. Sci. Fish. 6: 223-232.

NAUMAN, J. W. and R. L. CORY. 1969. Thermal additions and epifaunal organisms at Chalk Point, Maryland. Ches. Sci. 10: 218-226.

RUDDIMAN, W. F., D. S. TOLDERLUND, and A. W. H. BE. 1970. Foraminiferal evidence of a modern warming of the North Atlantic Ocean. Deep-Sea Res. 17: 141-153.

SANDERS, H. L. 1968. Marine benthic diversity: a comparative study. Am. Nat. 102: 243-282.

SCHOENER, A. 1974. Colonization curves for planar marine islands. Ecology 55: 818-827.

SHAW, W. N. 1967. Seasonal fouling and oyster setting on asbestos plates in Broad Creek, Talbot County, Maryland, 1963-1965. Ches. 8: 228-236.

WILHM, J. L. 1967. Comparison of some diversity indices applied to populations of benthic macroinvertebrates in a stream receiving organic wastes. J. Water Pollution Control Fed. 39 (10) Part I, 1673-1683.

WILHM, J. L. 1968. Use of biomass units in Shannon's formula. Ecology 49: 153-156.

WILHM, J. L. and T. C. DORRIS. 1966. Species diversity of benthic macroinvertebrates in a stream receiving domestic and oil refinery effluents. Am. Midl. Nat. 76: 427-449.

WOODS HOLE OCEANOGRAPHIC INSTITUTION. 1952. Marine Fouling and Its Prevention. U. S. Naval Inst., Annapolis, Maryland. 388 pp.

THE MECHANICAL EFFECTS OF WATER FLOW ON FISH EGGS AND LARVAE

Robert E. Ulanowicz
Center for Environmental and Estuarine Studies
Chesapeake Biological Laboratory
University of Maryland, Solomons, Maryland 20688

ABSTRACT

The impact of mechanical stresses upon ichthyoplankton entrained in power plant cooling systems has long been considered negligible. Arguments and evidence exist, however, to show that such a supposition is not universally true, especially in nuclear power plants. The mechanisms of mechanical damage can be detailed in terms of pressure change, acceleration, and shear stress within the fluid flow field. Laboratory efforts to quantify the effects of mechanical stress have been very sparse. A well-planned bioassay is urgently needed.

INTRODUCTION

In his pioneering paper on the effects of the Contra Costa electric generating facility on the local striped bass and salmon fisheries, Kerr (1953) devoted the bulk of his report to the mortality of adult fish due to impingement upon intake water screens. Passing mention was made of smaller juvenile forms passing through the screens to say that mortality due to mechanical stresses further along in the system was negligible. This work set the pattern for much of the research which followed. Most subsequent work on the theme of mechanical damage in cooling water systems repeated the theme that mechanical injury increased with increasing size (e.g., Markowski, 1962; Oglesby and Allee, 1969; Marcy, 1971).

Meanwhile, fish culturists were aware that mechanical trauma could be quite detrimental to the survival of juvenile fishes in hatcheries (Hayes, 1949; David, 1953; Leitritz, 1963). It is worth noting that much of the damage induced in developing young is not acute (i.e., immediate) but manifests itself later as deformities which could result in premature mortality. For example, Matlak (1970) inferred that misshapen heads in carp fry were attributable to mechanical disturbances. Emadi (1973) observed a high incidence of yolk sac deformation in alevins raised in trays with smooth bottoms--presumably a result of the fry's inability to maintain position and the consequent abrasion of the sac against the bottom surface. Mathur and Yazdani (1969) observed a vertebral mechanical compression in an adult Heteropneustes fossilis which he suggested was due to mechanical injury in early development.

If ichthyoplankton can be so sensitive to mechanical damage in the controlled environment of the hatchery, it would seem logical to expect some damage to the organisms when they are subjected to the rigors of the pumps and conduits of a cooling water system. Perhaps suspecting this, Barton Marcy (1973) was inspired to search for mechanical damage to entrained ichthyoplankton in the Connecticut Yankee Nuclear Power Plant. Marcy sampled the fish eggs and larvae in the discharge canal with the pumps operating, but without added heat or biocides. His data revealed that 80 percent of the entrained juveniles failed to survive the pass through the plant. While care must be taken not to transfer these results, quantitatively or qualitatively, to other power stations, they do nonetheless, point out the unequivocal need for more attention to the possibility of mechanical impact on entrained organisms.

Fortunately, an awareness of this problem is beginning to creep into environmental impact work. Lauer (1973) reported that pressure changes of the magnitude found in the Indian Point Power Plant were not damaging to striped bass eggs and larvae. Goodyear and Coutant (1973) added that one must also consider turbulence and shear as possibly inducing mortality. The authors of the impact statement for the Pilgrim Nuclear Power Station were careful to ascribe lobster larvae mortality to heat, mechanical action, and chemicals (A.E.C., 1972).

Mechanisms of Damage

While the whole issue of mechanical damage to entrained ichthyoplankton has not been studied thoroughly, it is possible, nonetheless, to outline the physical forces which could stress the organisms. It is important to emphasize that we are dealing with forces in the Newtonian sense. Thus, we should be careful not to speak of mortality caused by water velocity. The velocity, per se, of an organism or the fluid surrounding it can cause no damage. Rather, the three major forces (often associated with high velocities) to be reckoned with are (1) pressure change, (2) acceleration, and (3) shear.

The problem of pressure change, cause and effect, is the subject of another paper in this symposium, and hence will not be mentioned further in this manuscript.

Acceleration is the time rate of change of velocity and is always accompanied by a force according to Newton's second law of motion. In a cooling water system we can speak in terms of at least three ranges of acceleration.

At the lower end of the scale are the accelerations due to change in the bulk speed of the fluid flow. Typically these accelerations would be encountered near the intake as the water speeds up to its terminal velocity in the conduit, at the outlet as the water slows down, or at the upstream (or downstream) end of a constriction in the flow. These forces usually range from very slight to the order of magnitude of the gravitational force. In all likelihood they are not very damaging.

In the intermediate range are the accelerative forces associated with the turbulent eddies characteristic of most cooling water flows. If one follows the tortuous path of a particle in such a flow, he will become aware of abrupt changes in speed and direction. Such accelerations give rise to forces several times that of gravity and could possibly be damaging to ichthyoplankton, either immediately or in later development.

Potentially the most destructive accelerative forces would be short duration, high magnitude impulses from the impact of the organism with a solid surface. Such forces could be many times the acceleration of gravity and would probably be fatal.

Shear stress (or viscous force) is present in a real liquid when the velocity of the fluid varies from point to point in space. This is most commonly noticed when fluid moves with respect to a solid surface and exerts a viscous drag on the surface. The force is present throughout the fluid, however, and its potential effect on an egg in the flow field is illustrated in Figure 1. In this instance a fluid is flowing past a stationary surface. The fluid clinging to the surface wall has a velocity equal to 0. As one moves away from the wall the velocity increases up to some value characteristic of the bulk motion of the fluid. An egg within the changing velocity field would be subject to a fluid velocity on its outboard side (V') greater than that on its inboard side (V"). The resultant forces on the egg can be resolved (both conceptually and mathematically) into a rotational and a deformational component. The rotational effect would be to disturb the internal order of the egg, while the deformation would stress both the membrane and the interior.

The special case of damage incurred when the surface of the organism contacts the solid boundary is termed abrasion and deserves special study, since the resulting damage is a function of the two contacting surfaces.

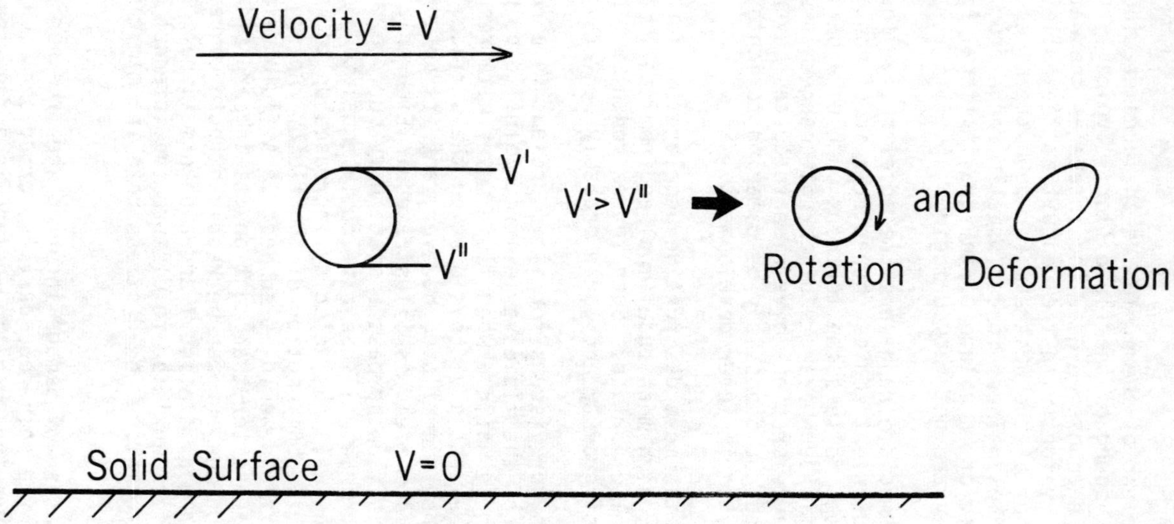

Figure 1. Possible effects of shear stress resulting from water velocity on a fish egg.

Shear stress is commonly reported by fluid dynamicists as a force per unit area (dynes/cm^2), however, the associated rate of shear (sec^{-1}) is often used in biological applications. For a Newtonian fluid, such as water, the two are directly proportional, the constant or proportionality being the coefficient of viscosity.

The destructiveness of shear is documented in the biophysical literature. It is a major cause of hemolysis in heart-lung machines (Williams, 1973; Bernstein, 1973), and is believed to be the principal agent in the degradation of fibrinogen within the circulatory system (Charm and Wong, 1970).

Within a flow field, the greatest shear stresses will be encountered in the near neighborhood of solid surfaces, e.g., conduit walls, pump impellers, screens, vanes, etc. In flows of high turbulent intensity one could also expect high shears to exist at the interfaces between eddies.

Previous Work

Now that the qualitative nature of possible mechanical damage to ichthyoplankton has been described, two questions arise: (1) What are the quantitative ranges of accelerative and shear forces which cause immediate and delayed damage to entrained fish eggs and larvae? (2) How does one design a cooling system to minimize such mechanical damage?

The engineering physics of flow systems is reasonably well understood. That is, a good chemical or mechanical engineer, given the design specifications of the system, should be able to invoke Boundary Layer Theory and the Theory of Turbulence along with empirically determined functions to specify where accelerations and shears of given intensities could be expected. His answers could never be determinant, since turbulent flows are stochastic by nature and certain complicated geometries can only be approximated, but his estimated should prove adequate to the task. Minimization of damage could be achieved by iteration of his calculations upon different design specifications.

The immediate bottleneck in evaluating the impact of entrained organisms is the paucity of bio-engineering data to answer the first question. In a reasonably intensive literature search, this writer has discovered only two attempts to gauge the effects of acceleration or shear upon fish eggs and larvae, neither of which was stimulated by power plant considerations.

Acceleration generated by shock waves from underground nuclear explosions was considered as possibly detrimental to trout redds in the vicinity of nuclear test sites (Post et al., 1973). Maximum peak accelerations in the order of 0.9 g were expected along one major trout stream. Post and associates, therefore, constructed a series of simulated rainbow trout redds in one-liter glass aquaria. An accelerometer was attached to replicate pairs of aquaria and the apparatus was dropped onto a resilient mat. Mean accelerations could be controlled to within 10 percent.

Pairs of aquaria were subjected to peak acceleration of 1, 2, 5, and 10 g's at biological development stages of 37.5, 75, 125, and 250 TU (one TU per hour is the product of the water temperature in degrees Fahrenheit above freezing divided by 24). Experiments were repeated at 8.3°C and 11.1°C. The survival rates of hatching (72-84 percent) did not differ significantly in any way from the experimental controls.

Extrapolation of these negative results should be done with caution. As the authors suggest, they probably apply to other salmonid species. But then, there are unconfirmed reports of researchers playing marbles with water-hardened salmon eggs! More fragile eggs, such as those of the striped bass, could prove to be more sensitive. Also, the larval stages of the stressed eggs should be examined for possible malformation or premature mortality. Finally, it would be useful to know what accelerations are necessary to induce mortality.

In an environmental impact study of the widening and deepening of the Chesapeake and Delaware Canal, Morgan et al. (1974) performed a series of simple experiments to assess the damage to striped bass, Morone saxatilis, and white perch, M. americana, eggs from shear fields.

Shear was generated by a plexiglass cylinder rotating axially in the middle of an annular tank of water. The experimental apparatus was designed for simplicity in fabrication, but without foreknowledge of the complicated flow which would be induced in the outer annular space. As a result, it became impossible to determine the exact value of shear which induced a given degree of mortality. It was possible, however, to calculate the maximum shear stress within the flow field and thereby provide reasonable values of shear to associate with observed impacts.

Eggs and larvae of both species were studied under long-term and short-term exposures. In the low-level, long-term experiments organisms were exposed to flows with characteristic shear rates ranging from 0.64 to 86.0 dynes/cm^2 for a period of two days.

Experiments with higher shear levels of from 76 to 404 dynes/cm^2 effective over exposure times of from 1 to 20 minutes completed the program.

Replication of the results of the experiments was very good. The lethal doses of shear required to kill half the specimens (LD_{50}) was calculated by polynomial regression from the experimental results and appears in Table 1.

The immediate question is how do the shear values in Table 1 compare with those within a typical nuclear power plant. Marcy (1973) reports for the Connecticut Yankee that at peak loads approximately 25 m^3/sec flows through twin, single-pass, divided water boxes in which the velocity may approach 2.4 m/sec. These figures imply a Reynolds' number for the flow in the order of 6 x 10^6. Fanning friction factors (see, Bird et al., 1960, p. 186) for such flow range from 0.0025 to 0.008 depending upon the smoothness of the surfaces of the conduit. Translated back into shear at the surface, stresses of from 72 to 230 dynes/cm^2 would be present at the walls of the water box.

While these values fall below the range of LD_{50}'s for one minute exposures (385-540 dynes/cm^2), they are within the damaging range. Higher shears could be expected in the pump, and it is probable that mortality in the experiment is being effected by shears lower than the wall stress. Therefore, the possibility that ichthyoplankton are being killed by shear stress in the nuclear cooling system is very real indeed.

SUMMARY AND RECOMMENDATIONS

Historically, the effects of mechanical stresses upon entrained ichthyoplankton have been assumed to be negligible. Mechanical damage to hatchery organisms is well documents, however. Nuclear power plants seem likely to inflict more mechanical damage to fish eggs and larvae passing through them because of their greater water use. The existence of a significant mechanical impact has already been demonstrated for the Connecticut Yankee Power Plant.

Existing laboratory research to quantify the damage to ichthyoplankton by acceleration and shear stress is sparse, fragmentary, and inconclusive. A well-coordinated and exhaustive effort is urgently needed.

Research is required to define the ranges of pressure change, acceleration, impaction, shear, and abrasion which cause mortality of ichthyoplankton of various entrained fishes and important inverte-

Table 1. Estimated LD_{50} values for time-shear exposure experiments on white perch and striped bass eggs and larvae. After Morgan et al. (1974).

Organism	Exposure Time	LD_{50} (dynes/cm^2)
Striped bass eggs	1 min.	450
	2 "	290
	4 "	170
	2 days	70
Striped bass larvae	1 min.	540
	2 "	435
	4 "	310
White perch eggs	1 "	385
	2 "	385
	5 "	150
	10 "	150
	20 "	150
	2 days	57
White perch larvae	1 min.	435
	2 "	402
	4 "	365
	2 days	88

brates. The determinations of damaging exposures should be made over all stages of development from fresh spawn to screenable juveniles. Temperature and salinity, of course, are important parameters to be varied in the experiments. Attention should be paid to the post-exposure development of survivors to assess possible delayed effects.

The huge quantities of heat added by power plants circumscribe the alternatives which would diminish thermal shock, but mechanical stress can be minimized by a large number of design alternatives. Such corrective action, however, awaits the development of reliable bio-engineering data.

ACKNOWLEDGMENTS

The author wishes to express his appreciation to Mrs. Carol Oen and Ms. Helen Pfuderer of the Ecological Sciences Information Center of the Oak Ridge National Laboratory for their assistance in performing a free automated information search. Thanks also go to Ray Morgan for citing several pertinent references.

LITERATURE CITED

ATOMIC ENERGY COMMISSION. 1972. Proposed issuance of an operating license to the Boston Edison Company for the Pilgrim Nuclear Power Station. Docket No. 50-293; Draft Environmental Impact Statement.

BERNSTEIN, E. F. 1973. Mechanisms of hemolysis. Annual summary report. Univ. of California at San Diego, La Jolla School of Medicine 406915.

BIRD, R. B., W. E. STEWART, and E. N. LIGHTFOOT. 1960. Transport phenomena. John Wiley and Sons Inc., New York.

CHARM, S. E., and B. L. WONG. 1970. Shear degradation of fibrinogen in the circulation. Science <u>170</u>: 466-468.

DAVIS, H. S. 1953. Culture and diseases of game fishes. Univ. of California Press, Berkeley.

EMADI, H. 1973. Yolk-sac malformation in Pacific salmon in relation to substrate, temperature, and water velocity. J. Fish. Res. Bd. Can. <u>30</u>: 1249-1250.

GOODYEAR, C. P., and C. C. COUTANT. 1973. Direct biological effects on entrained fish eggs and larvae at Indian Point.

Testimony to the Committee on Merchant Marine and Fisheries, U. S. House of Representatives, February 22, 1973.

HAYES, F. R. 1949. The growth, general chemistry and temperature relations of salmonid eggs. Quart. Rev. Biol. 24: 281.

KERR, J. E. 1953. Studies on fish preservation at Contra Costa steam plant of the Pacific Gas and Electric Company. Calif. Dept. Fish & Game Fish. Bull. 92.

LAUER, G. P. 1973. On the absence of mortality due to pressure changes on striped bass eggs and larvae. Testimony to the Committee on Merchant Marine and Fisheries, U. S. House of Representatives, February 5, 1973.

LEITRITZ, E. 1963. Trout and salmon culture. Calif. Dept. Fish & Game Fish. Bull. 107.

MARCY, B. C. 1971. Survival of young fish in the discharge canal of a nuclear power plant. J. Fish. Res. Bd. Can. 28: 1057-1060.

MARCY, B. C. 1973. Vulnerability and survival of young Connecticut River fish entrained at a nuclear power plant. J. Fish. Res. Bd. Can. 30(8): 1195-1203.

MARKOWSKI, S. 1962. Faunistic and ecological investigations in Cavendish Dock, Barrow-in-Furness. J. Anim. Ecol. 31: 43-51.

MATHUR, D. S., and G. M. YAZDANI. 1969. Observations on a deformed specimen of Heteropneustes fossilis Siluriformes Heteropneustidae. Sci. Cult. 35(9): 490-491.

MATLAK, O. 1970. Vorläufinger Bericht über Missbildung des Kopfes beim Karpfenstrich. Acta Hydrobiol. 12(4): 391-398.

MORGAN, R. P., R. E. ULANOWICZ, V. J. RASIN, L. A. NOE, and G. B. GRAY. 1974. Effects of shear on eggs and larvae of the striped bass, Morone saxatilis, and the white perch, M. americana. Submitted to Trans. Am. Fish. Soc.

OGLESBY, R. T. and D. J. ALLEE. 1969. Ecology of Cayuga Lake and the proposed Bell station (nuclear powered). Water Res. Mar. Sci. Cent., Cornell Univ., Ithaca, N.Y. No. 27: 315-328.

POST, G., D. V. POWER, and T. M. KLOPPEL. 1973. Effect of physical shock on developing salmonid eggs. Trans. Am. Nucl. Soc. 17: 27-28.

WILLIAMS, M. C. 1973. Shear-induced hemolysis: Materials and blood composition. Univ. California, Berkeley 071850.

ENTRAINMENT OF ORGANISMS AT POWER PLANTS, WITH EMPHASIS ON FISHES - AN OVERVIEW

Barton C. Marcy, Jr.
C. W. Rice Division, NUS Corp.
1910 Cochran Road, Pittsburgh, Pa. 15220

ABSTRACT

An overview of the recent information available on entrainment, specifically on fish eggs and larvae, is presented. The components of inner-plant mortality (thermal, mechanical and biocidal effects) are discussed, with emphasis on recent data on mechanical damage. Sampling considerations for entrainment studies as well as philosophies and suggestions for reducing entrainment losses are presented. Power plant impact can be minimized by siting in non-productive areas and going to closed cooling systems, i.e., lower intake volumes, since the mortality of meroplankton is directly related to the volume entering the intake. The low volume concept is presently the only effective approach to minimizing the adverse effects on entrained organisms. If studies continue to reveal high percentages of mechanical damage relative to thermal or chemical effects, future alternative strategies may also include the raising of condenser ΔT's while lowering the intake water volume, i.e., decreasing the number of entrained organisms vulnerable to mortality.

INTRODUCTION

Recently environmental studies of power plants have shifted emphasis from the effects of heated discharge to studies of the entire cooling system. Because of their relatively immobile, free-floating character, planktonic organisms are highly vulnerable to being "entrained" or passively drawn into the cooling water condenser systems of power plants. The importance of entrainment at water intakes is related to the species and the quantity of water withdrawn, the numbers of organisms in the receiving waters, their survival rates during entrainment as related to natural survival, their ecological roles, and their reproductive strategies. Factors affecting entrainment include the location of the power plant, the design of the cooling system, and the ambient receiving water conditions used for cooling.

Nonscreenable plankton or small fishes which pass through the intake screens intact are susceptible to various and simultaneous stresses, often leading to inner-plant mortality. Entrainment survival is determined by (1) the ambient temperature and the quality of the receiving water, (2) the seasonal densities, sizes, life-stages and relative susceptibility to injury of the species involved, (3) the amplitude of the temperature rise (ΔT) as the water passes through the condenser cooling system, (4) the duration of exposure to these temperatures, (5) the mechanical abrasions resulting from turbulence (shear forces) and pressure changes in passage through the system, (6) the exposure to biocides used for fouling control, and (7) gas bubble disease or the formation of air embolism caused by pressure and temperature changes in the cooling system.

It is the intent of this paper to provide an overview of recent information regarding the entrainment of nonscreenable organisms, specifically fish eggs and larvae at power plants; to discuss the components of inner-plant mortality with emphasis on mechanical damage; to provide a basis for entrainment sampling considerations; and to present suggestions for reducing entrainment losses.

Inner-Plant Mortalities

Overview

Losses of ichthyoplankton (fish eggs and larvae) at water intakes represent a larger impact on the ultimate population of adults over a broader geographical area than do the losses of phytoplankton and zooplankton. This is due to the longer regeneration time for fish, e.g., 2-4 years in the Patuxent River, Maryland (Mihursky, 1969). Most phytoplankton and zooplankton mortality is localized in the discharge plume or immediately downstream. Phytoplankton populations may be re-established to normal production within several hours after passage through a plant (Morgan and Stross, 1969). Talling (1962) has reported that under favorable conditions one to three cell divisions per day are typical of many freshwater phytoplankton species in the exponential phase of growth. During the warmer season, the minimum regeneration time for crustacean zooplankton (e.g., copepods) is about 5 days after passage through a plant (Mihursky, 1969).

There has been much emphasis in the past on field and laboratory studies of thermal tolerances and the thermal component causing inner-plant mortalities. Almost all of them have been concerned with phytoplankton and zooplankton. Generally the literature indicates that phytoplankton survive higher temperatures than zooplankton. In a recent review of the literature, the staff of the Atomic Energy Commission (1973a) found that losses of zooplankton in cooling systems ranged from 15 to 100 percent, but suggested that 30 percent "may be more representative." The limited data that are available concerning the passage of fish eggs and young through power plants show mortalities of between 39 and 100 percent, with the vast majority near 100 percent (Table 1).

Field and laboratory studies have been concentrated primarily on the lower levels of the food chain. As Clark and Brownell (1973) put it, ". . . it is far simpler to catch the tiny phytoplankton and the invertebrate zooplankton and to amass numerically significant data on them than it is to do the same for small fish . . . sampling the larval and early juvenile stages of fish is most difficult and requires advanced techniques and special expertise." As late as 1971, Marcy (1971) reported on the first attempt to determine survival rates of young fish of less than 15 mm. during passage through a power plant. Coutant (1971) in his review of 15 current entrainment studies at power plants, found only three being conducted on fish entrainment. The results of these were based mainly on preliminary field observations. Thus, we have seriously neglected the consumer level--many of the species entrained are important to man both recreationally and commercially, to say nothing of their ecological importance.

Thermal Effects

Up to 1972 most studies of inner-plant mortality (few of which were concerned with meroplankton (reproductive stages of fish and invertebrates which remain in the plankton for only a portion of their life cycle) have dealt only with the effects of increased condenser temperatures and have not separated the effects of biocide addition and mechanical abrasion within the cooling system. Marcy (1973) reported that, of the 100 percent mortality of the young of nine fish species passing through the Connecticut Yankee Atomic Plant, only 20 percent was attributable to heat, while 80 percent was due to mechanical abrasion during the 93 second passage through the cooling system. Schubel (1973) conducted laboratory experiments of time-temperature exposure histories of various developmental stages of fish eggs of some sensitive species, i.e., blueback herring, alewife, American shad, white perch and striped bass. He concluded that "our data clearly indicate that ΔT's of up to 10°C during the time of the year that these species spawn are not detrimental to egg development or hatching success." Also, Hoss et al. (1973) found that larval menhaden, spot, pinfish and three flounder species showed little or no distress when exposed to thermal shocks up to 12°C. A survey by Coutant (1970) of 61 power-reactor designs filed with the AEC revealed an average temperature rise across the condensers of 10.8°C. Thus, the temperature rise

Table 1. Inner-plant entrainment mortalities of fish eggs and young

Source	Location	Percentage Mortality
Marcy (1971, 1973)	Connecticut Yankee Plant, Haddam, Connecticut	100
Profitt (1969)	White River, Indiana	100
Clark and Brownell (1973)	Indian Point Plant, Hudson River, New York	97.5
Carpenter *et al*. (1972)	Millstone Point Plant, Niantic, Connecticut	High
EPA (1972)	Brayton Point Plant, Mount Hope Bay, Mass.	100
Lauer *et al*. (1974)	Indian Point Plant, New York	39
Flemer *et al*. (1971a)	Vienna Plant, Maryland	99.7
Flemer *et al*. (1971b)	Chalk Point Plant, Maryland	92.4
AEC (1973b)	Oyster Creek Plant, New Jersey	100

noted in some entrainment studies at power plants.

Species Entrained	Life Stage	Estimated Numbers Lost
Alewife, Blueback herring, White perch, Carp, White catfish, Spottail shiner, American shad, Johnny darter	Larvae + early juveniles 97.5% (2.6-15.0 mm) 2.5% (15.1-40 mm)	179 million/ year
Spottail shiners	Young	--
Not specified	Larvae and early juveniles	--
Flounder, Menhaden, Blueback herring	Larvae, juveniles	up to 20 million/day
Menhaden, River herrings	Larvae	7-165.5 million/day
Striped bass, Hogchoker, White perch, River herring	5 species of eggs 17 species of larvae	--
Striped bass	Eggs	--
Not specified	Larvae	--
24 species	Eggs and Larvae	150 million eggs/year 100 million larvae/year

experienced by entrained organisms at many plants may not in itself be the sole cause of inner-plant mortality, as once believed.

Survival of entrained fish will depend, in part, on the length of time the fish are exposed to the elevated temperatures. Duration of exposure in turn depends on the design of the cooling system and the length of the discharge area. Since little cooling takes place in a discharge canal (1-2°C) (March, 1971), the organisms are exposed to the elevated temperatures during the entire time they are entrained. The use of a long discharge canal can prolong the exposure to the maximum temperatures, and organisms are therefore exposed for a period equal to the flow time through the canal (Davies and Jensen, 1974). For example, Marcy (1971, 1973) has shown a 34 percent survival of entrained young fish (ΔT = 6°C) immediately after passage through the Connecticut Yankee plant, a 28 percent survival half-way down the 1.83 km canal and a 7.5 percent survival at the end of the canal. With a doubling of the ΔT to 12°C (full power output), the survival immediately after passage through the plant was 16.6 percent, or half that recorded at a ΔT of 6°C, and no living fish were taken at the end of the canal above a temperature of 28°C. At temperatures between 29° and 33.5°C, no fish survived passage half-way down the canal and at 35°C, 100 percent mortality occurred during passage through the plant. Marcy (1973) concluded that approximately 20 percent of the total mortality which occurred during plant passage was caused by heat shock and prolonged exposures (50-100 minutes) to temperatures above 28°C.

Davies and Jensen (1974) have concluded that "high velocity discharges into relatively large surface waters have been shown to produce temperature gradients that minimize the area of influence to discharge temperatures with obvious reductions in thermal stress to both pump and plume entrained plankton. By contrast, relatively low velocity discharges into smaller water bodies tend to increase the duration of exposure to water temperatures more closely approximating that of the condenser cooling system." Thus, shorter discharge canals with higher velocities will help to minimize mortalities due to heat shock and prolonged exposure to elevated temperatures.

Biocidal Effects

The injection of biocide, usually sodium hypochlorite, before the condensers to prevent fouling has caused mortalities of fish in the discharge at various plants. Marcy (1973) reported that the chlorination procedure at Connecticut Yankee is similar to that discussed by Draley (1972), in which the addition of chlorine before the water reaches the condensers and the splitting of the coolant stream into parts (chlorination of one bay at a time) allows for its removal in the same manner by reaction with slower (organic) chlorine demand constituents when remixing occurs. March (1973) found that biocide mortalities were insignificant because samples in the discharge involved only 25 percent of the total treated water volume and residual chlorine was recorded at less than 0.1 mg/l. Fairbanks et al. (1971) found that chlorine residuals at the 0.1 mg/l level had no adverse effects on menhaden, while ranges of 0.8 to 1.5 mg/l incapacitated them within minutes in a discharge canal. They also recommended treatment of each half of each condenser alternately. Split-stream injections and discharge chlorine residuals of less than 0.1 mg/l appear to cause minimum impact on fishes.

Mechanical Effects

The percentage of the total mortality caused by mechanical damage has not been adequately assessed at most power plants. Entrainment studies are just now beginning to separate the effects of mechanical, thermal and chemical stresses and to describe their synergistic effects. Among recent power plant studies, the following have shown mechanical damage to be the highest single cause of entrainment mortality: Marcy (1973) (Larvae of river herring, American shad, white perch, carp, white catfish, spottail shiner, johnny darter); Flemer et al. (1971a) (striped bass eggs); Edsall and Yocum (1972) (menhaden larvae); Tarzwell (1972) (menhaden larvae); Lauer et al. (1973) (eggs and larvae of river herring, hogchocker, white perch, striped bass); Schubel (1973) (blueback herring, striped bass, white perch and American shad eggs); Austin et al. (1973) (larval menhaden, weakfish, tautog, cunner); Carpenter et al. (1974) (zooplankton); Grosse Ile Laboratory, Inc. (1972) (zooplankton); Beck and Miller (1974) (zooplankton); and Flemer et al. (1971b) (phytoplankton). If studies continue to reveal high percentages of mechanical damage relative to thermal or chemical effects, future alternative strategies include the raising of condenser ΔT's while lowering the intake water volume, i.e., decreasing the number of entrained organisms vulnerable to mortality.

Fish, in particular, seem to be very vulnerable to mechanical stress. Marcy (1973) estimated that 80 percent of the 100 percent mortality of the young of nine fish species (2.6-40 mm) entrained at the Connecticut Yankee Plant was caused by mechanical damage. Flemer et al. (1971a) found extremely high mortality (99.7 percent) of striped bass eggs passing through the Vienna, Maryland, generating station and a high percentage of disintegration of the eggs. The death of 164.5 million menhaden larvae during one day at the

Brayton Point Plant (Rhode Island) was attributed to mechanical damage (Edsall and Yocum, 1972). Austin et al. (1973) reported 27-57 percent maceration (mechanical) mortality to four species of juvenile marine fish after plant passage at Northport, New York. Tarzwell (1972) reported that the only identifiable parts of larval menhaden (25-38 mm) after passage through a fossil-fuel plant were their eye balls. Of all studies reviewed, Lauer et al. (1973) reported the lowest (7-39 percent) mechanical damage mortality of entrained ichthyoplankton. Schubel (1973) stated that mechanical damage may have a much greater impact on development and hatching success than does temperature, i.e., ΔT's up to 10°C were not detrimental to egg development or hatching success of five estuarine fish.

Morgan et al. (1973) have measured the mortality caused by the shear fields (variation in water velocity with respect to time, expressed as units of force per unit area, i.e., dynes/cm^2) that are created by water movement over the surface of fish eggs and larvae. These experiments were performed on white perch and striped bass. The data obtained were used to predict the shear or mechanical stress created by the water passing through the Chalk Point fossil-fuel plant. Results indicated conservative estimates of mortality between 20 and 50 percent for both species (R. P. Morgan and R. E. Ulanowicz, personal communication). Morgan, in a paper presented at the Northeast Division of the American Fisheries Society Meeting, February 1974, calculated the shear or mechanical force of water passing through the Calvert Cliffs nuclear plant and estimated that shear stress alone would be enough to cause 50 to 100 percent mortalities of entrained eggs and larvae.

Two recent studies have shown much higher mortalities of zooplankton from the mechanical component than had been reported in earlier power plant studies. Carpenter et al. (1974) reported that 70 percent of the copepods entering the cooling water system of the Millstone Point Plant at Niantic, Connecticut, were killed by the mechanical or hydraulic stresses of passage. Beck and Miller (1974) cited unpublished data which demonstrated that mechanical damage caused 60 percent mortality of phytoplankton and zooplankton, with the primary contributor being pumping effects. Before entry into the pumps, the average mortality was 10 percent; after pumps, 50 percent; after condenser passage, 60 percent. Thus, the pumps may be a major contributor to mechanical damage.

Mortalities of between 29 and 55 percent for zooplankton passage at the Big Rock power plant on Lake Michigan were attributed to both mechanical and heat damage (Grosse Ile Laboratory, 1972). According to McNaught (1972), the greatest mortality of zooplankton passing through the Zion plant was caused by mechanical damage.

Initial findings of the effects of zooplankton passing through the condensers at the Waukegan plant showed an average mortality of 5.8 percent due to mechanical effects and 1.6 percent due to thermal stress (Industrial Bio-Test Laboratories, Inc., 1972). Studies of zooplankton at the Point Beach Nuclear plant showed that physical damage incurred by condenser passage was more critical than the thermal impact (8-19 percent mortality) Edsall and Yocum, 1972). A 5-20 percent loss in mobility of zooplankton was observed at many plants as the result of passage through just the pumps and condensers (Statement by Dr. Wright, Westinghouse Environmental Systems Department, in testimony at the Wisconsin hearings, Department of Natural Resources, 1971).

There appears to be a conflict of results on the effects of mechanical damage to phytoplankton during condenser passage. Gurtz and Weiss (1972) reported that some stimulation of photosynthesis occurred from the mechanical effects of condenser passage at the Allen generating plant on Lake Wylie, North Carolina. Williams et al. (1971) found no significant mechanical effects and Smith and Brooks (1971) found that some stimulation (17.5-30 percent) in productivity could occur from mechanical effects. On the other hand, Hardy (1971) observed an 11 percent reduction in productivity due to factors other than heat. Flemer et al. (1971b) found a 13 percent reduction in productivity which he attributed to mechanical effects. Morgan and Stross (1969) concluded that mechanical damage, cell disruption due to chlorination, or some other factor inhibited photosynthesis in samples of the effluent of the Chalk Point plant.

Different species and life-history stages also show different susceptibilities to mechanical damage. The high percentage of mechanical damage (80 percent) noted in the Connecticut Yankee study (Marcy, 1973) was linked to the fact that 97.5 percent of the entrained fish were in the critical post-yolk sac stage and 97 percent of the larvae were fragile clupeids. A high silt and detritus load passing through the condenser with the organisms may have been partially responsible for this high mortality by causing additional abrasion.

Increased mechanical injury with increased size during condenser passage has been reported for a few species of fish (Markowski, 1962; Oglesby and Allee, 1969; Marcy, 1971, 1973). Marcy (1973), for example, found that the greatest mechanical damage occurred at night when larger fish (20-40 mm) were available for entrainment. During the day the majority of fish were less than 15 mm long. McNaught (1972) found that mortality due to abrasion varied from 3.1 percent when small zooplankton were most abundant to 11 percent when larger zooplankton were most abundant. It is possible that mechanical damage during condenser passage could cause changes in the community structure by selecting against larger plankton.

Coutant (personal communication) found in the laboratory that mechanical damage may be the greatest cause of mortality to larvae of bluegill, carp and gizzard shad. The degree of damage was related to size, and most of the mechanical damage was due to pressure effects. Rapid pressure changes are characteristic of water passing through the cooling system, i.e., pressures fall just ahead of the pump impeller, build up to around 25 to 30 psi at the condenser tubes and then fall off to around 2 to 5 psi beyond the condenser tubes (Clark and Brownell, 1973). These pressure changes are particularly hazardous to the fish because of rapid expansion and contraction of the swim bladder; thus, larger fish with well developed swim bladders may be more susceptible to pressure effects than larval fish.

Nonscreenable entrained fish may also be susceptible to gas bubble disease, when under certain conditions the heated discharge water may be supersaturated with nitrogen and the gas that enters through the gills may bubble out of the fishes' blood streams. Again the effects are size dependent (DeMont and Miller, 1971), larger fish being more susceptible due to increase in body temperature (Marsh and Gorham, 1905) and stronger muscular activity (Egusa, 1959).

Sampling Considerations

The following are some sampling considerations for entrainment survival estimates:

1. The number of living eggs and larvae should be estimated in the net or pump samples in the receiving water that is committed to the plant's intake. The total percentage survival in the discharge is then calculated by dividing the total number of living larvae or eggs per cubic meter at the intake. A field observation device, used for maintaining water at ambient and at elevated temperatures during the counting of living and dead specimens, has been described by Marcy (1971). Dressel _et al_. (1972) have used vital stains to sort live and dead copepods.

2. The total numbers of organisms per cubic meter of water at the intake should be statistically comparable to those at the discharge side of the condensers. Often more samples are needed on the discharge side to obtain good confidence limits, as dead larvae settle out rapidly and fish distribution is changed by turbulent mixing, organism mobility, and changes in the specific gravity of the water in the discharge due to temperature changes.

3. Net or pump-caused mortalities must be kept to a minimum, and must be comparable in both the intake and discharge samples. Marcy (1973) found that in 0.3-0.6 m/sec current velocities at both intake and discharge sides, 5 minute stationary net collections with No. 2 mesh (0.39 mm) nets held mortality due to combined netting and natural causes to 18 percent. If velocities at the intake and discharge are variable, then measures should be taken to adjust volumes to obtain similar catch/velocity measurements (Lauer *et al*., 1974).

4. An entrainment sampling program should be statistically valid in order to establish confidence limits, sampling error and variance, etc. Marcy (1973) determined that triplicate 5 minute stationary samples at the surface, middle and bottom depths of each of four intake bays and at each of five stations along the 1.83 km. Connecticut Yankee discharge were sufficient for subsequent comparisons.

5. Physical measurements should include temperature, dissolved oxygen, nitrogen saturation (gas bubble disease), pressure measurements in the condenser system, shear forces, and turbidity and detritus load passing through the system with the organisms. Sampling immediately before and after pumps and condensers will help to pinpoint mechanical stresses.

6. Developmental stages and sizes should be recorded during all determinations of the components of inner-plant mortality for estimates of relative vulnerability. It is suggested that rose bengal dye be added to a 3 percent neutral formalin solution to facilitate rapid separation and counting in the laboratory.

7. Diurnal (24 hour) studies are necessary to determine species and size-range vulnerability to the intake both day and night. For example, Marcy (1973) found fish larvae to be more abundant near the bottom during the day and between mid-water and surface at night near a plant's intake, thus being more vulnerable at night than during the day.

8. Sampling to determine the components of entrainment mortality (mechanical, thermal, biocidal) should be carried out over the entire spawning season of the potentially vulnerable fish species, at least every two weeks and weekly during peak egg and larval abundance. For example,

the Michigan Water Resources Commission advocates in their proposed guidelines that intake sampling for eggs and larvae be conducted one day per week for a 24 hour period as a minimum during one calendar year.

9. In estuaries, an organism that drifts with currents may be swept alternately upstream and downstream past a plant's intake as tidal flows change from ebb to flood tide. These organisms, therefore, may be vulnerable to entrainment several times before they pass downstream farther than the distance of one tidal excursion from the intake. Thus, entrainment studies must take tidal flux into account. Marcy (1973) reported the highest percentage of larvae were taken into the Connecticut Yankee plant's intake on a slack-flood as opposed to ebb tide. He calculated the number of nonscreenable living fish passing by the plant and the percentage of those that were entrained. The average entrainment was 4 percent (95 percent confidence limits, 3.4-4.5) under conditions of unidirectional net tidal flow:

$$\frac{(N_i)\ (F_i)}{(N_r)\ (F_r)}\ (100) = \%$$

where N_i = average number of living fish per cubic meter in the intake cross section (3 depths x 4 intake bays); F_i = flow rate into intake in m^3/sec; N_r = average number of living fish per cubic meter in river cross section (3 depths, 50 m apart x 11 stations across river); and F_r = average net tidal flow rate (m^3/sec) in the river near the plant during each time period sampled. Carlson and McCann (1969) estimated that 4 percent of the eggs and 12 percent of the larvae of striped bass were subject to withdrawal from a river cross section near the intake of a pumped storage plant.

In standing bodies of water, the best sampling approach is probably a cubic design with appropriate length, width and depth intervals to determine densities of organisms in a defined cubic area in the vicinity of the intake. Assessments can then be made of the annual loss of numbers and biomass to the intake by relating the number of organisms per cubic meter in the receiving water with the volume entering the intake per unit of time.

10. Cooperation with biologists and plant engineers is essential to determine the percentages of mechanical, thermal and biocide damage. For example, mechanical damage should be determined with all intake pumps running at capacity

with no addition of heat or biocide. Biocide damage should be estimated with all pumps running, with no heat added and with the normal biocide procedure initiated. Thermal damage should be estimated at a variety of plant heat loads along the length of the discharge to determine upper temperature limits and time-temperature survival curves.

CONCLUSIONS

Ecologists have spent many years refining classic techniques and increasing the reliability of sampling and identification methods. We are now faced with decisions on total ecological effects and impacts when present "state-of-the-art" knowledge is inadequate. We are asked by engineers and politicians to develop cost-benefit estimates. For example, of the millions of larval fish killed by entrainment in power plant cooling systems each year, how many would have survived to become adults and what is their dollar value (say versus the cost of a cooling tower)? Using these narrow criteria the organism will always lose. Since entrainment mortality is an additive above natural mortality, it may affect ultimate populations of adults, community structure and the total ecological balance. We need to know the natural variation in fish populations in order to assess the impact of entrainment losses. We need to know the natural survival rates between life stages, the carrying capacity of the system, and the total dynamics of a fishery in order to assess the impact of a power plant, or more importantly, the cumulative effects of several power plants on a water body.

Modeling can be utilized to predict the impact of proposed plants, or modifications of existing plants. Through modeling we can begin to utilize the data gathered in pre-operational studies rather than to merely list the data collected. Model studies can, therefore, provide a vehicle for bringing the results of environmental impact studies into perspective for use in environmental legislation, i.e., to provide a pre-development picture of possible post-development impacts. One approach is simple population modeling. For example, one may determine the fecundity of a species, and assign a percentage mortality between life stages through actual experiments or by assuming a species which exhibits high parental care and carries low numbers of eggs has a higher survival between the egg and larval stage than the reverse case. Then by estimating the numbers lost to the intake per year and assuming one female must produce two fish to maintain a stable population, one can estimate the potential population of adults lost.

It would seem impossible to screen small fish larvae effectively. It has been found, for example, that impingement of striped bass eggs and larvae on fine mesh screening for even a short time is fatal. There are at least two types of intake screens now being tested in the United States to screen ichthyoplankton. However, no data are presently available.

Coordinating plant activities with organism densities is one approach which can help to alleviate the entrainment problem: planning plant shutdowns, for refueling or maintenance, to coincide with high peaks of egg and larval density; varying pumping rates, day vs. night or flood vs. ebb tides, corresponding to known densities; and possibly using multiple intakes in different areas or depths so as to draw water from areas of low fish densities.

Power plants can minimize impact by siting in non-productive areas and in many cases by going to closed cooling systems, i.e., lower intake volumes, since the mortality of meroplankton is directly related to the volume entering the intake. This low volume concept is presently the only immediately effective approach to minimize the adverse effects on entrained organisms. Also it may be possible to increase condenser ΔT's while lowering intake volumes, especially since the mechanical damage component of the mortality appears to dominate.

We must also shorten the time between quantitative re-evaluations of data and the present necessity of making broad ecological "guess-estimate" decisions based on limited population dynamics data. Ecologists and engineers must work together to design cooling-water systems which minimize the impact of entrainment losses.

LITERATURE CITED

ATOMIC ENERGY COMMISSION. 1973a. Final environmental statement--Shoreham nuclear power station.

ATOMIC ENERGY COMMISSION. 1973b. Draft environmental statement related to the Oyster Creek nuclear generating station. Jersey Central Power and Light Company. Docket Number 50-219.

AUSTIN, H. M., J. DICKINSON, and C. HICKEY. 1973. An ecological study of ichthyoplankton at the Northport power station, Long Island, New York. New York Ocean Sci. Lab. Montauk, N.Y. Prepared for Long Island Lighting Co.

BECK, A. D., and D. C. MILLER. 1974. Analysis of inner plant passage of estuarine biota. Paper presented at the Amer. Soc. Civil Eng. Power Div. Spec. Conf., Boulder, Colo. Aug. 12-14.

CARLSON, F. T., and J. A. McCANN. 1969. Hudson River fisheries investigations (1965-1968), Rep. to Consolidated Edison Co., N.Y. by the Hudson River Policy Comm.

CARPENTER, E. J., S. J. ANDERSON, and B. B. PECK. 1972. Entrainment of marine plankton through Millstone Unit 1. 3rd and 4th semi-annual rep., Woods Hole Oceanographic Instit.

CARPENTER, E. J., B. B. PECK, and S. J. ANDERSON. 1974. Survival of copepods passing through a nuclear power station on northeastern Long Island Sound, U.S.A. Mar. Biol. 24: 49-55.

CLARK, J., and W. BROWNELL. 1973. Electric power plants in the coastal zone: environmental issues. Spec. Publ. No. 7, Amer. Litt. Soc., Highlands, N.J.

COUTANT, C. C. 1970. Biological aspects of thermal pollution. I. Entrainment and discharge canal effects. C.R.C. Crit. Rev. Envir. Control 1(3): 341-381.

COUTANT, C. C. 1971. Effects on organisms of entrainment in cooling water: steps toward predictability. Nuc. Saf. 12: 600-607.

DAVIES, R. M., and L. O. JENSEN. 1974. Effect of entrainment of zooplankton at three mid-Atlantic power plants. Elec. Power Res. Instit., Cooling Water Discharge Res. Proj. (RP-49), Publ. No. 74-049-00-1, Rep. No. 16, The Johns Hopkins Univ., 1.

DEMONT, D. J. and R. W. MILLER. 1971. First reported incidence of gas-bubble disease in the heated effluent of a steam generating station. Paper presented at the 25th Annual S.E. Assoc. Game Fish Comm., Oct. 17-20, Charleston, S.C.

DRALEY, J. E. 1972. The treatment of cooling waters with chlorine. Argonne Nat. Lab., Argonne, Ill., Envir. Proj. No. ANL/El-12, 12 p.

DRESSEL, D. W., D. R. HEINLE, and M. C. GROTE. 1972. Vital staining to sort live and dead copepods. Chesapeake Sci. 13: 190-193.

EDSALL, T. A., and T. G. YOCUM. 1972. Review of recent technical information concerning the adverse effects of once-through cooling on Lake Michigan. Prepared for the Lake Michigan Enforcement Conf., Sept. 19-21, 1972, Chicago, Ill. U.S. Fish Wildl. Serv., Great Lakes Fish. Lab., Ann Arbor, Mich.

EGUSA, S. 1959. The gas disease of fish due to excess of nitrogen. (In Japanese, Eng. Summ.) J. Fac. Fish. Anim. Husb. 2: 157-183.

ENVIRONMENTAL PROTECTION AGENCY. 1972. Proceedings in the matter of pollution of Mount Hope Bay and its tributaries.

FAIRBANKS, R. B., S. W. COLLINGS, and W. T. SIDES. 1971. An assessment of the effects of electrical power generation on marine resources in the Cape Cod canal. Mass. Dept. Nat. Res., Div. Mar. Fish.

FLEMER, D. A. 1971a. The effects of steam electric station operation on entrained organisms. In J. A. Mihursky and A. J. McErlean (co-principal investigators), Postoperative assessment of the effects of estuarine power plants. Nat. Res. Instit., Ref. No. 17-2b.

FLEMER, D. A. 1971b. Preliminary report on the effects of steam elecyric station operations on entrained organisms. A progress report to the Maryland Dept. Nat. Res. under the postoperative assessment of effects of estuarine power plants grant. In J. A. Mihursky and A. J. McErlean (co-principal investigators), Postoperative assessment of the effects of estuarine power plants. Chesapeake Bay Lab. Ref. No. 71-24a.

GROSSE ILE LABORATORY. 1972. Lake Michigan entrainment studies, Big Rock nuclear power plant, Escanabe power plant, Nov.-Dec. 1971. EPA Office of Research and Monitoring, Grosse Ile, Mich., Rep. No. 1.

GURTZ, M. E., and C. M. WEISS. 1972. Field investigations of the response of phytoplankton to thermal stress. Envir. Study Program, Duke Power Co., Univ. of N.C., ESE Publ. No. 321.

HARDY, C. D. 1971. Dissolved oxygen survey at Northport, p. 116-119. In G. C. Williams, Studies on the effects of a steam-electric generating plant on the marine environment at Northport, New York. Tech. Rep. No. 9, Marine Sci. Res. Cent., State Univ. New York, Stony Brook, N.Y.

HOSS, D. E., W. F. HETTLER, JR., and L. C. COSTON. 1973. Effects of thermal shock on larval estuarine fish-ecological implications with respect to entrainment in power plant cooling systems. Paper presented at the Proceedings of the Symposium on the Early Life History of Fish, Oban, Scotland.

INDUSTRIAL BIO-TEST LABORATORIES, INC. 1972. Intake-discharge experiments at Waukegan generating station. Rep. to Commonwealth Edison Co., Proj. XI, IBT No. W9861, Biol. Sec.

LAUER, G. J. 1973. Entrainment studies on Hudson River organisms. New York Univ. Medical Center, Lab. Envir. Stud., November.

LAUER, G. J., W. J. WALLER, and D. W. BATH. 1974. Fish egg and larvae entrainment by the Indian Point Power Plant on the Hudson River estuary. Paper presented at the N.E. Div. Amer. Fish. Soc. Meeting, February.

MARCY, B. D., JR. 1971. Survival of young fish in the discharge canal of a nuclear power plant. J. Fish. Res. Bd. Canada 28: 1057-1060.

MARCH, B. D., JR. 1973. Vulnerability and survival of young Connecticut River fish entrained at a nuclear power plant. J. Fish. Res. Bd. Canada 30(8): 1195-1203.

MARKOWDKI, S. 1972. Faunistic and ecological investigations in Cavendish Dock, Barrow-In-Furness. J. Anim. Ecol. 31: 43-51.

MARSH, M. C. and F. P. GORHAM. 1905. The gas disease in fish. Rep. U. S. Bur. Fish., 345-376.

McNAUGHT, D. C. 1972. The potential effects of condenser passage on the entrained zooplankton at Zion Station. Prepared statement at the Lake Michigan Enforcement Conf., Sept. 21, 1972, Chicago, Ill.

MIHURSKY, J. A. 1969. Patuxent thermal studies - summary and recommendations. Univ. of Maryland Nat. Res. Instit., Spec. Rep. No. 1.

MORGAN, R. P., III and R. G. STROSS. 1969. Destruction of phytoplankton in the cooling water supply of a steam electric station. Chesapeake Sci. 10(3-4): 165-171.

MORGAN, R. P., III, R. E. ULANOWICZ, V. J. RASSIN, L. A. NOE, and G. B. GRAY. 1973. Effects of water movement on eggs and larvae of striped bass and white perch. Nat. Res. Instit.,

Chesapeake Biol. Lab., Univ. of Maryland, NRI Ref. No. 73-111.

OGLESBY, R. T. and D. J. ALLEE. 1969. Ecology of Cayuga Lake and the proposed Bell station (nuclear powered). Water Res. Mar. Sci. Cent., Cornell Univ., Ithaca, N.Y., No. 27: 315-328.

PROFITT, M. A. 1969. Effects of heated discharge upon aquatic resources of the White River at Petersburg, Indiana. Indiana Univ. Water Res. Cent., Rep. of Invest. No. 3.

SCHUBEL, J. R. 1973. Effects of exposure to time-excess temperature histories typically experienced at power plants on the hatching success of fish eggs. Chesapeake Bay Instit., The Johns Hopkins Univ. Spec. Rep. No. 32-PPRP-4, Ref. No. 73.11.

SMITH, R. A. and A. S. BROOKS. 1971. Primary production, p. 32-37. In R. W. Koss (ed.), An interim report on environmental responses to thermal discharges from Marshall steam station, Lake Norman, North Carolina. Cooling Water Discharge Proj. (RP-49), Edison Elec. Instit., The Johns Hopkins Univ., Baltimore.

TALLING, T. F. 1972. Freshwater algae. In R. A. Lewin (ed.), Physiology and biochemistry of algae. Academic Press, N.Y.

TARZWELL, C. M. 1972. An argument for open ocean siting of coastal thermal electric plants. J. Envir. Qual. 1(1): 89-91.

WILLIAMS, G. C. 1971. Studies on the effects of a steam-electric generating plant on the marine environment at Northport, New York. Tech. Rep. No. 9, Mar. Sci. Res. Cent. State Univ. New York, Stony Brook, N.Y.

THE ASSESSMENT OF IMPACT DUE TO ENTRAINMENT OF ICHTHYOPLANKTON

Thomas J. Horst
Ecologist, Environmental Engineering Division
Stone & Webster Engineering Corp.
Boston, Massachusetts 02107

ABSTRACT

Three approaches to the assessment of impact associated with the entrainment of ichthyoplankton into power stations are discussed. The first is the estimation of the probability of mortality for those organisms entrained. The second is the assessment to the fish populations which have the egg and larvae life stages entrained. The third approach is the assessment of impact to the community or ecosystem. Some models are presented as examples for each of the three approaches.

INTRODUCTION

This paper presents a perspective on the state of the art for assessing the impact to aquatic ecosystems as a result of entraining ichthyoplankton into the cooling water systems of power stations. Some ideas which show promise for impact assessment in the near future are outlined.

Three general approaches, the individual, the population and the community, each corresponding to a different level of biological organization, are presented for the evaluation of impact. Most studies to date have focused on the individual; however, the population approach is now being used in a few studies. The first approach concerns the probability of survival for an organism which has been entrained into the circulating water system of a power station. The second approach concerns the effect on a species population which incurs mortality due to entrainment of its early life stages. The last approach, termed the system approach, considers the effect to a community of organisms which incurs mortality due to entrainment. The types of parameters necessary for each approach are discussed and the difficulties in obtaining site specific data are explored.

The ultimate decision concerning the acceptability of any man-induced impact such as entrainment is a human value judgement. It is essential therefore, to provide information gathered from a sound biological basis and statistical rationale in order to make the most responsible value judgements possible. The most useful data are those gathered with a view to present ecological theory and in the realization of the limitations of the present state of the art of field studies.

THE ORGANISM APPROACH

Impact may be assessed for those individuals which experience entrainment, and consequently the probability that the individual will die as a result of passage through the cooling water system can be determined. Many studies have focused on this question and several of the papers contained in this symposium are examples of such work (Schubel, 1975; Ulanowicz, 1975; Beck, et al., 1975; Bongers and Polger, 1975; and Marcy, 1975). The mortality associated with passage through the cooling water systems may be associated with increased temperature, pressure, chemicals, and mechanical effects. Since these factors operate simultaneously in the power station, laboratory experiments which are designed to simulate a single factor do not generally give the information necessary to estimate the probability of an organism surviving entrainment. The simultaneous simulation in the laboratory of all factors encountered during entrainment is exceedingly difficult, if not technically impossible.

An alternative approach has been to conduct field studies which sample the organisms entering the intake and those organisms contained in the discharge. The ratio of live to dead organisms is then determined at both locations. In this type of field study the density of organisms at the intake (Ni) should theoretically be the same as the density at the discharge (Nd).

$$Ni = Nd \tag{1}$$

If this is the case, the reduction in the percentage of live organisms from the intake (Pi) to the discharge (Pd) represents the mortality due to passage through the circulating water system (M).

$$M = Pi - Pd \tag{2}$$

The delayed mortality associated with entrainment must then be determined by holding a sample of organisms which have passed through the power station and collected at the discharge. A control sample must also be monitored to estimate the mortality associated with the holding process.

Generally the mean value for mortality is high, and the confidence limits for the estimates are large for fish eggs and larvae passing through open cycle cooling water systems (Marcy, 1973) and for copepods (Carpenter *et al.*, 1974). The upper confidence limit at the 95 percent level is often above the maximum 100 percent mortality. The precision of these estimates may be increased. However, considerable increase in sample size and consequently cost will be necessary. A more rigorous experimental design will be necessary to reduce the presently unexplained variability of these estimates.

For closed cycle cooling systems where the residence time of an organism in the cooling system is much longer and the mechanical turbulence is greater than for once through cooling, the probability of death to an entrained organism increases to almost a certainty. For closed cycle cooling power stations therefore, it may be most practical to simply assume total mortality of those ichthyoplankers entrained.

THE POPULATION APPROACH

Once the probability of death associated with passage through the power station has been determined, the questions of how many organisms are lost from the population due to entrainment and the effect of this loss on the species population need to be considered.

Entrainment may be viewed as an additional source of mortality to the early life stages (eggs and larvae) of the fish population. The fish population can often be defined by a population model and the parameters of the model estimated from field data or literature values. Most population models do not consider spatial distribution of the individual organism, but assume the probability of death for all organisms at a particular age to be the same. Since entrainment is a point source mortality the evaluation of the entrainment mortality must consider spatial distribution of the organisms, and the physical dynamics of the water body.

The choice must be made either to simulate the population dynamics on a temporal basis in one model and the entrainment events both spatially and temporally in another model; or to simulate both events with spatial and temporal components in one model. Depending on the hydrodynamics of the particular situation, development of a single model for both the entrainment events and the population dynamics can be a difficult task both in mathematical formulation and in terms of computer simulation.

Simulation of the Hudson River striped bass population as discussed by Lawler (1972) is an example of the single model technique. The winter flounder model presented by Hess, Sissenwine and Saila (1975) is an example of the two model technique.

Saila, Sissenwine and Hess (1974) have shown that the two model approach produces reductions in the population levels due to entrainment mortality equal to or greater than those of the single model method. It is therefore a more conservative technique since the bias is to over-estimate the impact of entrainment. The two model method will be explored in more detail in the following discussion.

The simulation of the entrainment mortality should address the question of the probability of an organism in the population being entrained into the intake. This estimate can be made in several ways, but essentially it becomes the ratio of the number of organisms entrained to the total population of the organisms. When only the eggs and larvae of a fish are entrained, the total population for each of these life stages must be estimated. This is a difficult problem since the geographic boundries for population, the development time for each life stage, and any other species specific characteristics must be considered. The models used for the entrainment estimates are physical models based on ideas such as mass balance. The physical model presented by Hess, Sissenwine and Saila (1975) in this symposium serves as an example of this type of model.

The modeling of the population dynamics should be done with a view to the limitations of the available data to parameterize the model. Most of the population models presently being used are variations on the life table approach. Populations for which the age specific mortalities and fecundities can be estimated may be simulated by a computerization of the Leslie (1945) model. This model is:

$$\vec{N}_{t+1} = \underline{A}\,\vec{N}_t \qquad (3)$$

where $\vec{N}$ is a 1 by x vector corresponding to the number of organisms (n_i) in each of x life stages, and each life stage has an equal development time.

$$\vec{N}_t = \begin{bmatrix} n_1 \\ \cdot \\ \cdot \\ \cdot \\ \cdot \\ n_x \end{bmatrix}_t \qquad (4)$$

The x by x $\underline{A}$ matrix is the projection matrix which describes the transition of the population from time t to t+1.

$$\underline{A} = \begin{bmatrix} F_1 & F_2 & \text{----------} & F_{x-1} & F_x \\ P_1 & 0 & \text{----------} & 0 & 0 \\ 0 & P_2 & \text{----------} & 0 & 0 \\ 0 & 0 & \text{----------} & 0 & 0 \\ \vdots & \vdots & \ddots & \vdots & \vdots \\ 0 & 0 & \text{----------} & P_{x-1} & 0 \end{bmatrix} \qquad (5)$$

where

F_i is the number of female offspring born to a female of age i

P_i is the probability that an organism of age x will survive to age x+1.

and

$$P_i = e^{-d_i} \qquad (6)$$

where

d_i is the instantaneous death rate of organisms of age i.

The finite rate of population growth (R) can be calculated from the $\underline{A}$ matrix as the maximum characteristic root of the characteristic equation with the stable age distribution represented as the characteristic vector accompanying the largest root (Leslie, 1945).

$$R = \text{max char root } \underline{A} \qquad (7)$$

The instantaneous population growth rate (r) is:

$$r = \log_e R \qquad (8)$$

One method for investigating the impact due to entrainment is to estimate the elements of the $\underline{A}$ matrix from field and literature values for the affected population. The value of r can then be calculated. The P_i or the probabilities of survival to the next age can then be converted to instantaneous age specific death rates. The instantaneous mortality rate due to entrainment can be added to the age specific death rate and the new instantaneous death rates converted to a new age specific probability of survivorship. The maximum characteristic root and the associated characteristic vector of the second A matrix represent the instantaneous population growth rate and the stable age distribution of the impacted population.

The instantaneous rates of population growth with and without entrainment can be compared as can the stable age distributions. This permits comparisons of the theoretical potential of the population under the assumptions of exponential growth, and is therefore most conservative as there is no density dependence in the population. A computer program EIGENPOP was developed by the writer to numerically solve the A matrix for the characteristic roots and vector. This program is based on the EISPAC routines contained in Reinsch and Wilkinson (1971).

The analysis of the year to year variation in population size may be made by simulating the population represented by the Leslie (1945) model. A computer program called POP1 was also developed which simulates the Leslie model. This program calculates the probability of survivorship to the next age as the combination of the instantaneous rates of natural, fishing and entrainment mortality. Any of the elements in the A matrix can be constants or functions of the population density. The population is then simulated with and without the effect of the power station. The change in population size or any selected population parameter represents the impact associated with entrainment by the power station of fish eggs and larvae.

This life table approach using the Leslie model can be made stochastic by using the formulations of Chiang (1960a,b). If the life table is assumed to represent a cohort of organisms the variance associated with the chance of an organism of age x dying before it reaches age x+1 (Qi) has been shown to be:

$$Qi = \frac{(1 - Qi)\ Qi}{n_i} \left[1 + \frac{1}{n_i} - \frac{1}{n_1} \right] \qquad (x = 1,2,3\text{-----}4) \tag{9}$$

where

n_i is the number of individuals of age i

and

$$\hat{V}(\hat{P}_i) = \hat{V}(1-\hat{P}_i) = \hat{V}(\hat{Q}_i) \tag{10}$$

Chiang (1960a,b) has also shown that the covariance between estimates of Qi in a cohort are equal to zero

$$\text{cov}\ (Q_i,\ Q_j) = 0 \qquad (i \neq j) \tag{11}$$

Once the variances of the elements of the $\underline{A}$ matrix are computed and some distribution function is assumed for the Q_i, a stochastic simulation of the population can be made. A computerization of this model called POP2, which assumes the elements of the $\underline{A}$ matrix are normally distributed, was developed.

For some populations it may not be possible to make all the estimates for the elements of the $\underline{A}$ matrix. It may be more useful to use a stock and recruitment relationship and the mortalities through the year classes. Although the stock and recruitment function has many variations, the Ricker (1958) model serves as an excellent example of this type of model:

$$F = P e^{P_r - P/P_m} \qquad (12)$$

where F is the number of recruits at some stage after density dependence ceases in the population.

P is the parental stock

P_r is the replacement size of the parental stock

P_m is the level of the parental stock which produces the maximum number of recruits.

This function is difficult to parameterize since it requires a long time series of both stock and recruitment estimates and also requires data over a wide range in values for both the stock and the recruits. Results of Waller et al. (1971) are an example of the least squares parameterization, while the results of Hess, Sissenwine and Saila (1975) are an example of a deterministic parameterization of the stock and recruitment function. The simulation of the population using a stock and recruitment function and the Leslie model are the same once the population has been defined.

The population methods discussed thus far require a great deal of information for the affected population. Because of the nature of these parameters it is difficult to estimate them from field studies, and many times they do not exist in the literature. A simplistic approach is to translate the number of ichthyoplankers lost to entrainment into the number of adults that would have resulted assuming no compensatory mechanisms in the population.

If the population is in equilibrium, in one generation the fecundity produced by a breeding pair will be reduced to two breeding adults

$$2 = S \cdot F \tag{13}$$

where S is the survival from egg to adult

F is the fecundity of a breeding pair during their life,

or $$S = 2/F \tag{14}$$

If the entrained life stage is larvae, then the survival from egg to larvae (S_e) must be multiplied times F to give the survivorship from larvae to adult (S_l).

$$S_l = \frac{S}{S_e} = \frac{2}{S_e F} \tag{15}$$

The number of entrained larvae (N_l) is multiplied by S_l to give the number of adults (N_a) that would have resulted assuming no density dependence.

$$N_a = S_l N_l \tag{16}$$

The number of adults can then be compared to some reference such as catch statistics for commercial or sport species. This may be a meaningful comparison when sufficient information is not available for the more extensive analysis.

THE SYSTEM APPROACH

The impact of a power station on an aquatic community or ecosystem may also be assessed. Since it is not practical to consider each species in the entire ecosystem, major components which adequately represent the structure and function of the system both before the power station went into operation and after must be considered. Defining the ecosystem becomes one of the most difficult tasks in this approach.

The models used to represent communities or ecosystems have typically been compartment type models which consider the flow of energy and material transfer through the system. The most important organisms are typically those which represent the major pathways of energy and material transfer. In some situations the most important organisms may not be the major groups affected by power plant-induced mortality.

The systems approach will not be explored in detail in this paper since present field studies and impact assessment have not used this approach extensively. A change in the direction of studies to this approach would require more rigorously defined field studies and more sophisticated modeling programs.

The same general types of analyses which were discussed for the population approach can be extended for community analysis. Patten (1971 & 1972) has assembled an excellent collection of papers on simulation using the systems approach. Since the mechanisms of community dynamics are not presently as clearly defined as population dynamics, there are few widely used models for community analysis. Most often the differential equations for the compartments are derived empirically from the available data, and these differential equations are solved numerically.

Another approach to community analysis involves the community matrix ($\underline{A}$). It contains elements α_{ij} where, the α_{ij} is the relative amount by which species i will depress species j. The α_{ij} can be defined from the Lotka-Volterra equation:

$$\frac{dN_i}{dt} = (r_i)_0 \, N_i \left[\frac{1-N_i}{K_i} - \alpha_{ji} \frac{1}{K_i} N_j \right] \qquad (17)$$

for any two species i and j where

N_i is the number of individuals of species i

r_i is the intrinsic rate of increase of species i

K_i is the carrying capacity of the environment for species i.

Authors such as Levins (1968) have attempted to define α_{ij} in a more operational manner, however the collection of field data for this parameter is still one of the major difficulties. The work by May (1973) provides some interesting new ideas on the concept of community stability using the community matrix.

If the mortalities incurred by organisms as a result of power station operation can be translated into changes in the elements of the community matrix, these techniques may prove useful for impact analysis. This remains as a great gap between ecological theory and data gathered from applied field studies.

DISCUSSION

The three approaches discussed represent a sizable expense and effort for assessing impact. It seems justified to seek a pathway which yields the necessary data, but does not incur unneeded and overly expensive studies. There is a delicate balance between the cost of ecological studies, and the benefit in terms of answering the necessary questions.

The estimation of the probability of death for passage through a power station or the first approach should be carefully scrutinized. If the initial results indicate the mortalities are nearly 100 percent, or the confidence limits for the estimate are very large, it may be more prudent to make the initial impact assessment assuming 100 percent mortality. For closed cycle cooling with longer residence times and greater turbulence, the 100 percent mortality estimate may be realistic for fish eggs and larvae.

At present the population approach offers the best promise for impact assessment, because models are available which are generally accepted and the impact can be assessed in terms of a basic biological unit, the species population. The choice of the species for which impact is assessed should depend not only upon which species have the greatest losses due to entrainment, but also upon their importance to the ecosystem structure and function.

Since the ecosystem approach does not presently offer clear direction for impact assessment, the choice of species populations to be studied must depend upon the important components of the ecosystem. In this manner at least an indirect attempt is made to consider the impact on the whole ecosystem. The choice of the species must also be influenced by the data available for modeling the dynamics of these species.

ACKNOWLEDGMENTS

I would like to thank Drs. Paul Eisele and Peter Hofmann of the Environmental Engineering Division, Stone & Webster Engineering Corp., for their ideas during the development of this paper and their review of the manuscript. I would also like to thank Ms. M. Jackson, who typed the manuscript.

LITERATURE CITED

BECK, A. P., G. V. POJE and W. T. WALLER. 1975. A laboratory study on the effects of the exposure of some entrainable Hudson River biota to the hydrostatic pressure regimes calculated for Storm King. This symposium.

BONGERS, L. H. and T. T. POLGER. 1975. Effects of thermal and chlorine stresses on entrained estuarine zooplankton. This symposium.

CARPENTER, E. J., B. B. PECK and S. J. ANDERSON. 1974. Survival of copepods passing through a nuclear power station on northeastern Long Island Sound. Marine Biology 24: 49-55.

CHIANG, C. L. 1960a. A stochastic study of the life table and its applications. I. Probability distributions of the biometric functions. Biometrics 16: 618-635.

CHIANG, C. L. 1960b. A stochastic study of the life table and its applications. II. Sample variance of the observed expectation of life and other biometric functions. Human Biol. 32: 221-238.

HESS, K. W., M. P. SESSENWINE and S. B. SAILA. 1975. The impact of the entrainment of winter flounder larvae. This symposium.

LAWLER, J. P. 1972. Effect of entrainment and impingement at Indian Point on the population of the Hudson River striped bass. Testimony before U.S.A.E.C. Docket No. 50-247. Oct. 30, 1972.

LESLIE, P. H. 1945. On the use of matrices in certain population mathematics. Biometrika. 33: 183-212.

LEVINS, R. 1968. Evolution in changing environments: Some theoretical explorations. Princeton University Press, Princeton.

MARCY, B. C. 1975. Vulnerability and survival of young fishes entrained at power plant intakes. This symposium.

MARCY, B. C. Jr. 1973. Vulnerability and survival of young Connecticut River fish entrained at a nuclear power plant. J. Fish. Res. Board Can. 30: 1195-1203.

MAY, R. M. 1973. Stability and complexity in model ecosystems. Princeton University Press, Princeton.

PATTEN, B. C., ed. 1971. Systems analysis and simulation in ecology. Volume I. Academic Press, New York.

PATTEN, B. C., ed. 1972. Systems analysis and simulation in ecology. Volume II. Academic Press, New York.

REINSCH, J. W. and J. WILKINSON. 1971. Handbook for automatic computations. Volume I: Linear Algebra. Springer Verlag., Berlin.

RICKER, W. E. 1958. Handbook of computations for biological statistics of fish populations. Bull. Fish. Res. Bd. Can. No. 119.

SAILA, S. B., M. P. SISSENWINE and K. W. HESS. 1974. Evaluating the effect of power plant entrainment on populations near Millstone Point, Connecticut. MES-NUSCo Rep. No. 1. Northeast Utilities Service Company, Hartford.

SCHUBEL, J. R. and T. S. Y. KOO. 1975. Thermal effects of power plants on fish eggs and larvae: An assessment. This symposium.

ULANOWICZ, R. E. 1975. The mechanical effects of water flow on fish eggs and larvae. This symposium.

WALLACE, D. 1975. Critical comparisons of assumptions in striped bass models with field survey data. This symposium.

WALLER, W. T., M. L. DAHLBERG, R. E. SPARKS and J. CAIRNS JR. 1971. A computer simulation of the effects of superimposed mortality due to pollutants on populations of fathead minnows (Pimephales promelas). J. Fish. Res. Bd. Can. 28: 1107-1112.

LABORATORY AND FIELD TEMPERATURE PREFERENCE AND AVOIDANCE DATA OF FISH RELATED TO THE ESTABLISHMENT OF STANDARDS

J. R. Stauffer, D. S. Cherry, K. L. Dickson
and J. Cairns, Jr.
Biology Department
Virginia Polytechnic Institute and State University
Blacksburg, Virginia 24061

ABSTRACT

Temperature preferences for "important" fish species in the New River in the vicinity of Appalachian Power Company's Glen Lyn, Virginia plant were determined independently by both field and laboratory studies. A relationship was demonstrated between the temperature preference data generated by the two approaches. Based on the temperature preference data the responses of fish to the thermal discharges can be predicted. From these data and from other data on the fish community structure, it was possible to determine that the thermal discharge was causing no appreciable harm to the fish community. Based on these studies we feel that the most reasonable approach to establishing thermal standards is to couple temperature preference studies with site specific studies.

INTRODUCTION

The electric utilities presently use 70 to 80 percent of all water taken in by industry for cooling purposes (Berkowitiz and Squires, 1969; Parker and Krenkel, 1969). Parker and Krenkel (1969) further estimate that by the year 2000 "the equivalent of the total flow of surface water in the U.S. will be needed for cooling purposes if once through cooling is required." The concern about heated waste discharges from electric utilities is further intensified since the electric power demand in the United States has been doubling every 6 to 10 years (Cairns, 1972). In light of the growing problem of thermal pollution the Environmental Protection Agency (1973) has recently listed several proposed criteria for important species on which thermal standards will be based: 1) ultimate upper incipient lethal temperature; 2) temperature cold shock; 3) optimum temperatures for growth and reproduction; 4) maximum temperatures for short term exposures. However, for fishes, the EPA did not consider temperature preference and avoidance data in establishing thermal standards. In situations such as Appalachian Power Company's Glen Lyn, Virginia plant, located in the New River where the total heated discharge (528 cfs, 14.95 m^3/sec) is small compared to the average annual flow of the river (6,100 cfs, 141.33 m^3/sec) we feel that laboratory and field determined upper temperature avoidance and temperature preference data on important species might be an appropriate way to establish thermal standards. This would be especially true if laboratory temperature preference experiments reflect the temperature response of populations *in situ* as confirmed by comprehensive field studies. If the adult and fingerling fish have alternative water temperatures, they can avoid heated areas during the short intervals during which the discharge water may exceed their upper incipient lethal temperature. Obviously, this is not possible for larval fishes and other organisms which may not be able to avoid thermal discharges (i.e., entrained organisms) as noted by Coutant (1970, 1972).

Other authors have recognized the importance of temperature preference and avoidance studies, and have tested several species in laboratories (Doudoroff, 1938; Fisher and Elson, 1950; Fry, 1947; Brett, 1952; Garside and Tait, 1958; Meldrim and Gift, 1971). A comprehensive list of such references, and others on related material has been reported by Raney and Menzel (1969) and Raney, Menzel and Weller (1973). Although laboratory studies provide useful information about the temperature preference and avoidance of fishes they usually do not simultaneously evaluate other physical, chemical, and biological parameters which affect fish distribution.

Field studies of the temperature selection of fresh-water fishes are rare (Ferguson, 1958). Those studies which used field data for this purpose were limited mostly to reservoir and lake studies in which temperature varied with depth (Fry, 1947; Dendy, 1948; Kennedy, 1941). Ferguson (1958) also attempted to relate laboratory preference data to field data. However, a fairly comprehensive literature search did not identify studies which combined both on-site laboratory and field data in the evaluation of temperature response of fishes.

The purpose of this study was to use temperature responses for important species as determined by on-site field and laboratory studies, to evaluate the impact of a heated effluent and to use this information to determine whether the 81°F stream standard which is now in existence is appropriate.

RESEARCH SITE

Field

Heated waste water (i.e., above ambient temperature) entered the New River from two Appalachian Power Company (APCo) discharges at Glen Lyn, Virginia. One unit discharged 189 cfs (5.3 m^3/sec) directly into the New River approximately 50 yards (45.7 m) above the Rt. 460 Bridge. The second unit discharged 269 cfs (7.6 m^3/sec), via a 300 foot (274.3 m) long effluent channel, into the East River (a tributary of the New River) upstream from its confluence with the New River, and 68 cfs (1.9 m^3/sec) into the New River. Water from both units remained along the left bank (facing downstream) of the New River.

The New River varied in flow from a high of 82,800 cfs (2345 m^3/sec) recorded in May to a low of 1375 cfs (39 m^3/sec) recorded in September 1973. Flow data were not available for the East River. Field sampling stations, all in Giles County, Virginia, are shown in Figure 1. Stations were selected for accessibility, proximity to heated discharges, substrate composition, and presence of riffles and pools. The substrate at all locations consisted of different amounts of silt, gravel, rubble and bed rock.

Due to the location of the sampling stations at various places in and out of the thermal plume a variety of temperatures was present for each collection date. Thus fish had the opportunity to select their preferred temperature, or to avoid high temperatures.

Laboratory

The Glen Lyn research facility was a 60 ft. (18.2 m), mobile field laboratory located on the confines of the power plant, approximately 10 ft (3 m) from the edge of the New River (Fig. 2). The laboratory consisted of three compartments, two areas for holding facilities (three 250 gallon [946.2 l] holding tanks and several 50 gallon [189.2 l] aquaria) and a central area for testing fish responses. The central laboratory section included preference and avoidance troughs, circulating water baths, continuous temperature monitors, camera and closed-circuit television.

MATERIALS AND METHODS

Field Samples

Rotenone was used in conjunction with a block net to sample both pool and riffle habitats at each station (Hocutt *et al.*, 1973) from May 1972 to October 1974. A particular water temperature was delimited by the block net and then rotenoned. All fish were sorted, identified to species, and placed in permanent storage in the VPI&SU fish museum. Collection temperatures ranged from 69° to 96°F (20.5° to 35.5°C) and were measured in degrees F (most temperature standards [state and federal] are expressed in F) with a Taylor hand thermometer.

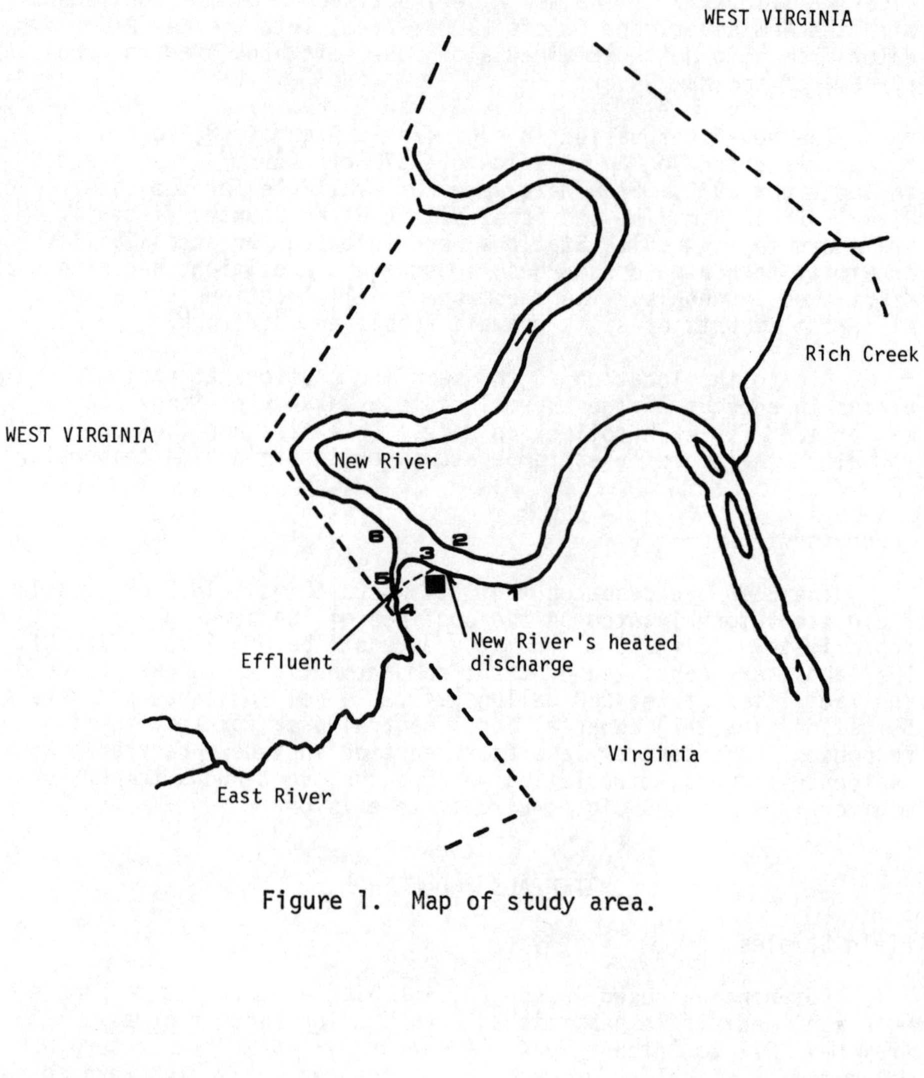

Figure 1. Map of study area.

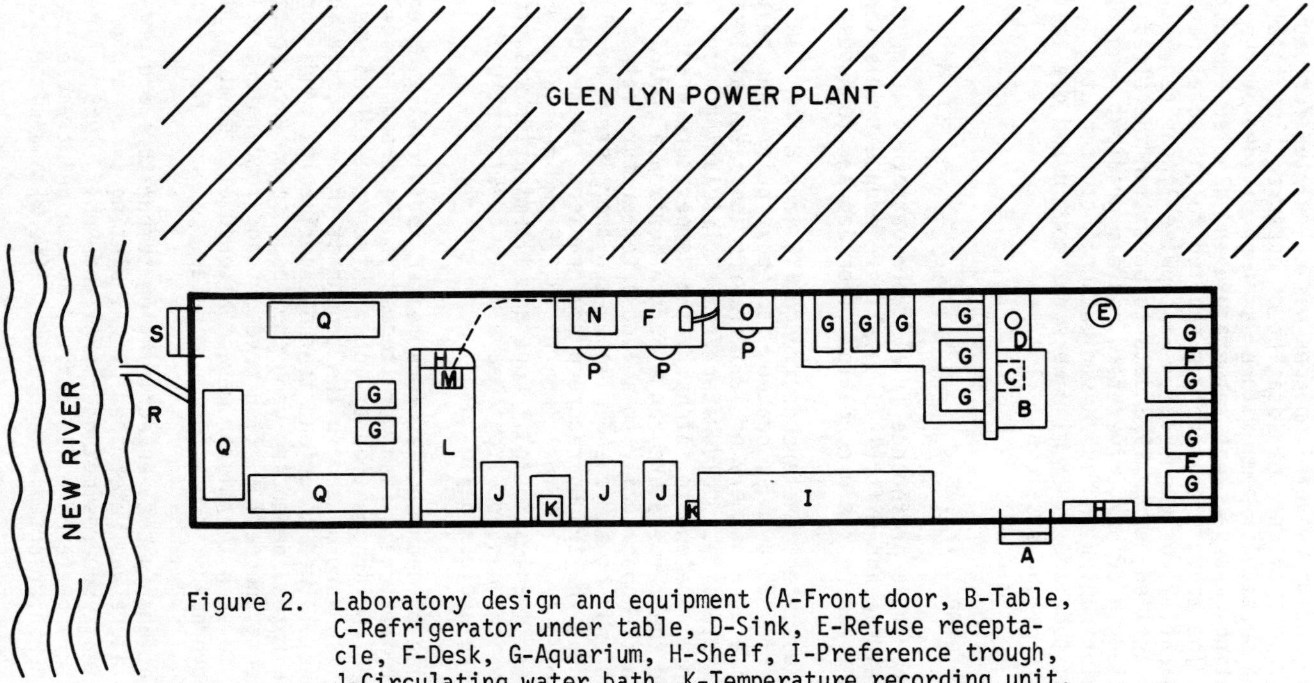

Figure 2. Laboratory design and equipment (A-Front door, B-Table, C-Refrigerator under table, D-Sink, E-Refuse receptacle, F-Desk, G-Aquarium, H-Shelf, I-Preference trough, J-Circulating water bath, K-Temperature recording unit, L-Avoidance trough, M-Television camera, N-Closed circuit TV monitor, O-Computer teletype monitor and telephone coupler system, P-chair, Q-Holding tank, R-Water intake pipes from immersion pump in river, S-Back door).

The abundance of several species was calculated for each temperature regardless of date or location of capture. The total abundance at a particular temperature was then divided by the number of collections made at that temperature and plotted against temperature. The above procedure was applied to the total data base, and separately for each year. A stepwise linear regression technique was applied to the 1973 data to partition the effects of temperature, length of day, river flow, stream gradient, and time since last chlorination on fish distribution (Stauffer et al., with editors). Altitude and river mile were not considered due to the close proximity of the stations.

Laboratory Preference Trials

The temperature preference trials were patterned after those reported by Meldrim and Gift (1971). All preference trials were conducted under rising field conditions from January to August 1974.

A stainless steel preference trough (13 in x 8 in. x 10 in. [33 cm x 20.3 cm x 25.4 cm]), coated with epoxy paint, was heated on the underside by a battery of 12 infra-red lamps set at increasing levels of intensity. Cold water from a circulating water bath was introduced at one end to intensify the temperature gradient. This increased the temperature range of the horizontal gradient to allow the fish the opportunity to select a temperature between undesirable cold and hot extremes. Twenty-two thermistors equally spaced throughout the trough recorded the gradient. Water depth in the trough varied from 1/2 to 1 inch (1 to 2.5 cm) depending upon the size of each species tested. Constant illumination was provided by three incandescent lights at a level consistent with solar radiation during a clear day.

Eight fish of each species were tested individually at each acclimation temperature. Fish were tested in the trough prior to gradient formation for position effects. After a gradient was established, fish were placed in the trough at their acclimation temperature and allowed to orient to the experimental conditions for 20 to 60 minutes, depending upon the response of each species. After the orientation period, fish movement and temperature selection relative to each thermistor were monitored at 1 minute intervals through overhead mirrors for 20 minutes.

The relationship between acclimation temperature and preference temperature was determined by calculating a regression line via the least squares method (Sokal and Rohlf, 1969). The final temperature preference was defined as the point where the regression line intersected a theoretical line with a slope of 1, according to Fry (1947).

Species for all preference trials were collected by seine from the New River drainage, stored in holding tanks and then transferred to one of several aquaria. Holding facilities were equipped with immersion heaters for acclimation at a constant temperature for one week. Acclimation levels tested ranged from 53.6°F to 96.8°F (12.0°C - 36.0°C) in 5.4°F (3°C) increments. Food was withheld from fish for 2.5 to 3 days prior to a trial in order to maintain a consistency in stomach contents. Differences in fork length (FL) within a species was minimal (0.9 in. [2.28 cm]), but differences in FL between species varied from 1.9 in. to 5.8 in. (4.82 cm - 14.73 cm). In all tests, fish were used once and then discarded.

RESULTS

A total of 28,432 specimens represented by 43 species were collected via 25 rotenone collections from May 1972 to October 1974 at temperatures which ranged from 67°F to 96°F (19.4°C - 35.5°C). Since rotenone is temperature sensitive, the collection of fishes at temperatures cooler than 60°F (15.5°C) was prohibited. The laboratory tests resulted in the establishment of temperature preferendum for 25 species at rising and falling acclimation levels ranging from 42.8°F to 96.8°F (6°C - 36°C) depending upon the species. From the above data a total of seven species were selected according to the EPA (1973) criteria. The EPA (1973) defined important species as: 1) important commercial or sport fishes, 2) abundant forage fishes, and 3) rare and endangered species. Of the 43 species collected in this study, none was considered either commercial or rare and endangered.

The spotfin shiner (Notropis spilopterus, Cope) was selected as an important forage fish. It was the most abundant species in 7 of the 25 rotenone collections. As can be seen by Figure 3, its distribution appeared to be unaffected by temperature. In 1973 it was equally abundant at 69°F (20.5°C) and 95°F (35°C), while in 1974 it was most abundant at 80°F (26.6°C). The laboratory data indicated that TP = .47 A + 45.37 where TP is equal to the temperature preference and A is equal to the acclimation temperature ($P \leq .01$, df = 70) with 82 percent of the data explained by the regression equation. Cherry *et al.* (1974) reported an upper avoidance of 95°F (35°C) of spotfin shiners when acclimated to 86°F (30°C). It is interesting to note that all of the fish acclimated at 96°F (35.5°C) selected a temperature below the predicted value.

The stoneroller (Campostoma anomalum, Rafinesque) was selected as another important forage fish. It was the most dominant species in 7 of the 25 rotenone collections. The field data indicate that the stoneroller definitely preferred temperatures cooler than 75°F (23.8°C).

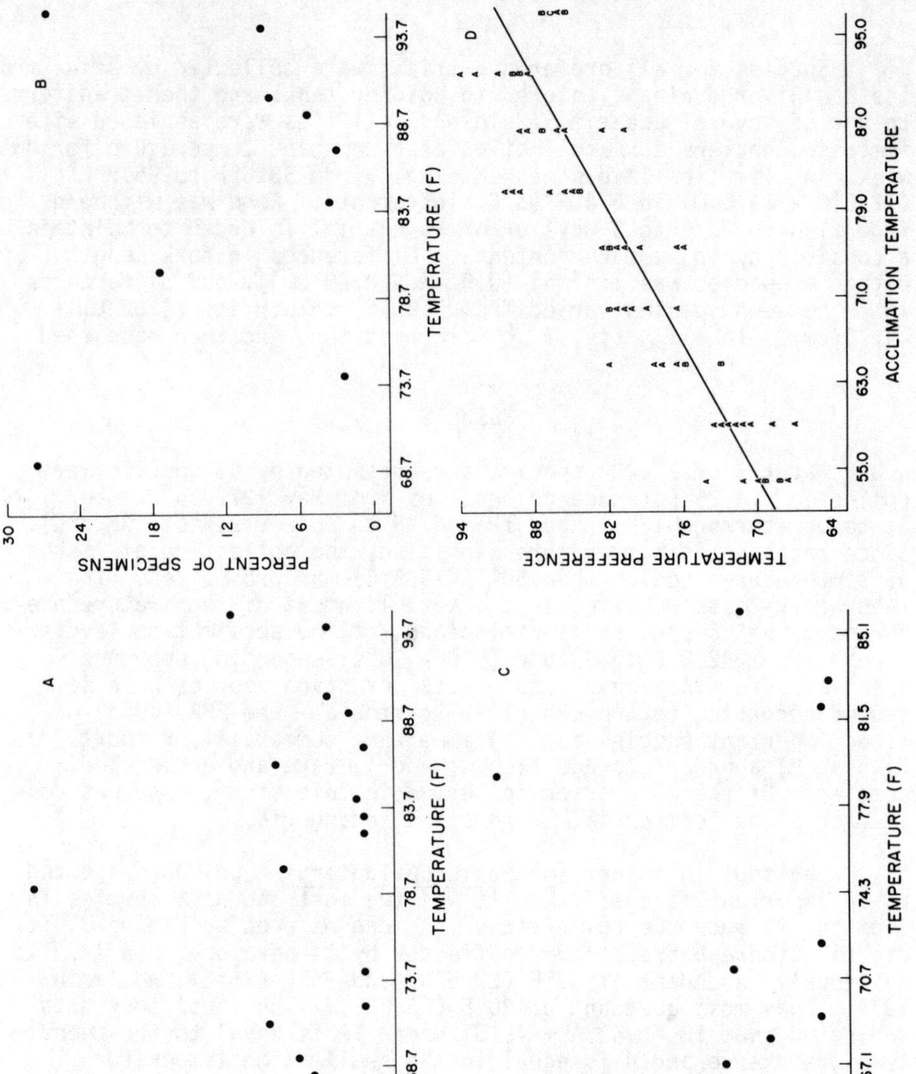

Figure 3. Temperature responses of the spotfin shiner (<u>Notropis spilopterus</u>, Cope). A - Relative abundance (corrected for sampling effort) of composite data for 1973 and 1974; B - 1973 data; C - 1974 data; D - Laboratory relationship between acclimation temperature and preference temperature, where A = 1 observation, B = 2 observations, etc.

In 1973, more than 90 percent of the specimens were captured at temperatures cooler than 79°F (23.8°C); and in 1974, more than 95 percent of the specimens were recorded in water temperatures cooler than 73°F (22.7°C) (Fig. 4). The laboratory preference data are defined by the equation TP = .71A + 23.3 ($P \leq .01$, df = 54); with 94 percent of the data explained by the regression. The final temperature preference as defined by Fry (1947) was 80.34°F (26.8°C).

The northern hogsucker (*Hypentelium nigricans*, Lesueur) was selected both because it is an important forage species and because it is known to migrate into streams to spawn (Breeder and Rosen, 1966). Therefore, the thermally influenced portion of the East River may have been a block for migrating fish. Both the 1973 and 1974 field observations show a gradual increase in abundance as temperatures rise from 67°F to 79°F (19.4°C to 26.1°C) (Fig. 5). The laboratory data was explained by the regression, TP = .27A + 60 ($P \leq .01$, df = 38), with 44 percent of the data explained by the regression (Fig. 5). The calculated ultimate preference temperature is 82.2°F (27.8°C). Cherry *et al*. (1974) reported an upper avoidance temperature of 95°F (35°C) when fish were acclimated to 86°F (30°C).

The channel catfish (*Ictalurus punctatus*, Rafinesque) was selected as representing the most important sport fish in the area (Bryson *et al*., in press). During the 1973 season greater than 90 percent of the specimens were captured in water exceeding 94°F (34.4°C) (Fig. 6). The data for 1974 are not as definitive; however, this represents a very small percentage of the total catch as can be seen when all of the specimens are grouped (Fig. 6A). The final temperature preference calculated from Cherry's *et al*. (1974) data was 92.82°F (33.7°C).

Two other potential sport fish were the spotted bass (*Micropterus punctulatus*, Rafinesque) and the smallmouth bass (*Micropterus dolomieui*, Lacepede). Too few specimens of these species were collected to determine any temperature relationships from the field data. Furthermore, Hatch (1973) reported movement of largemouth bass (*Micropterus salmoides*, Lacepede) into a heated discharge as a response to the presence of forage fish. The laboratory data for the smallmouth bass were best explained by the regression P = .39A + 53.42 ($P \leq .01$, df = 46) (Fig. 7A) with 72 percent of the data explained by the regression line. The calculated ultimate temperature preference was 83°F (28.3°C). The regression line which best explained the data for the spotted bass was TP = .31A + 61.8 ($P \leq .01$, df = 46 (Fig. 7B) with 62 percent of the data explained by the regression. The calculated final temperature preference was 86°F (30°C).

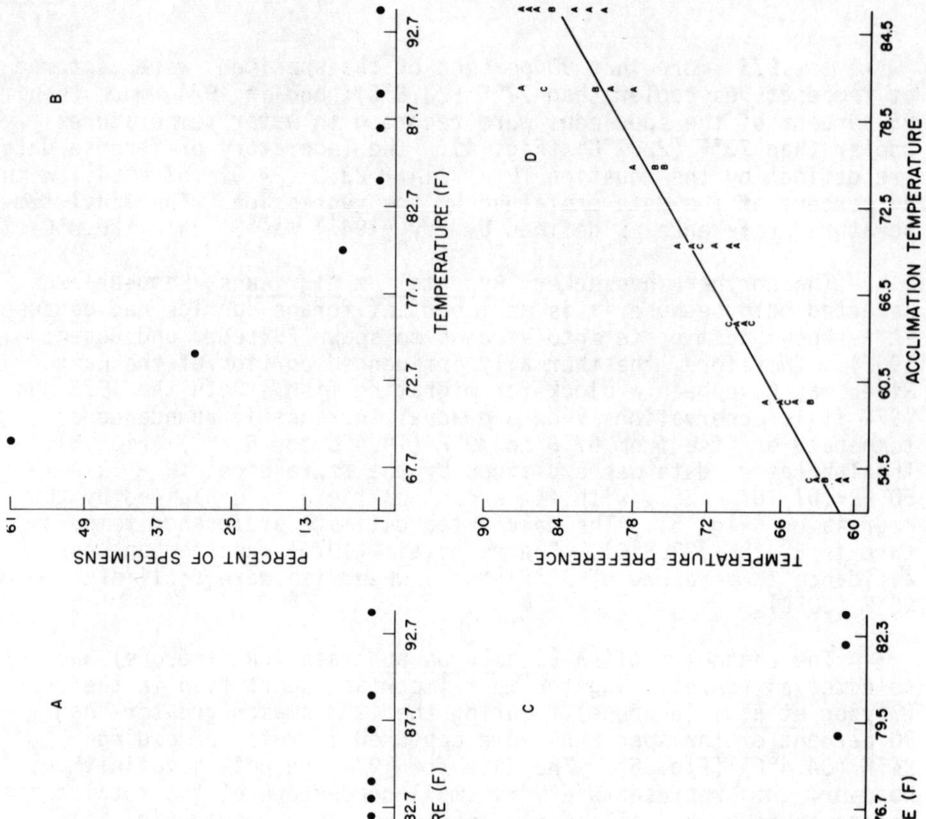

Figure 4. Temperature responses of the stoneroller (Campostoma anomalum, Rafinesque). A - Relative abundance (corrected for sampling effort) of composite data for 1973 and 1974: B - 1973 data; C - 1974 data; D - Laboratory relationship between acclimation temperature and preference temperature, where A = 1 observation, B = 2 observations, etc.

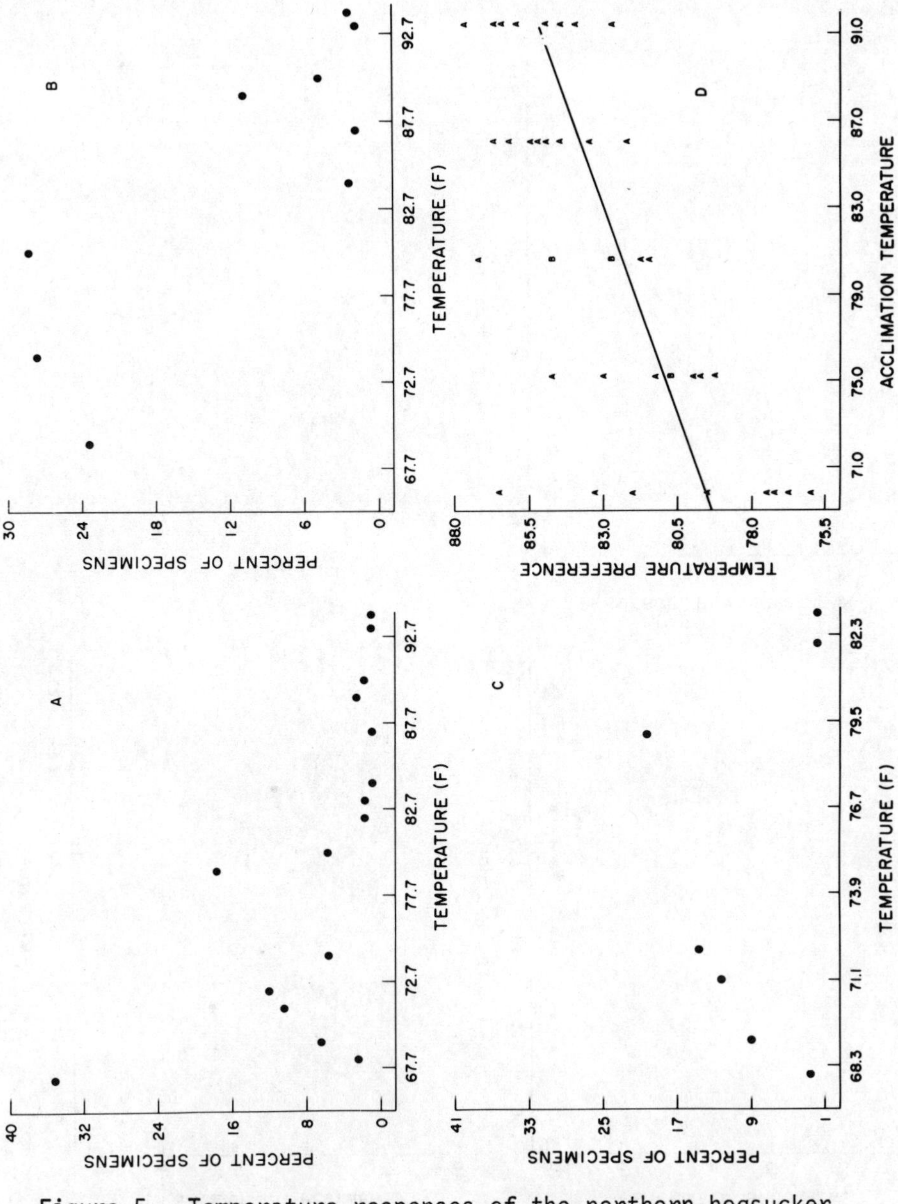

Figure 5. Temperature responses of the northern hogsucker (Hypentelium nigricans, Lesueur). A - Relative abundance (corrected for sampling effort) of composite data for 1973 and 1974; B - 1973 data; C - 1974 data; D - Laboratory relationship between acclimation temperature and preference temperature, where A = 1 observation, B = 2 observations, etc.

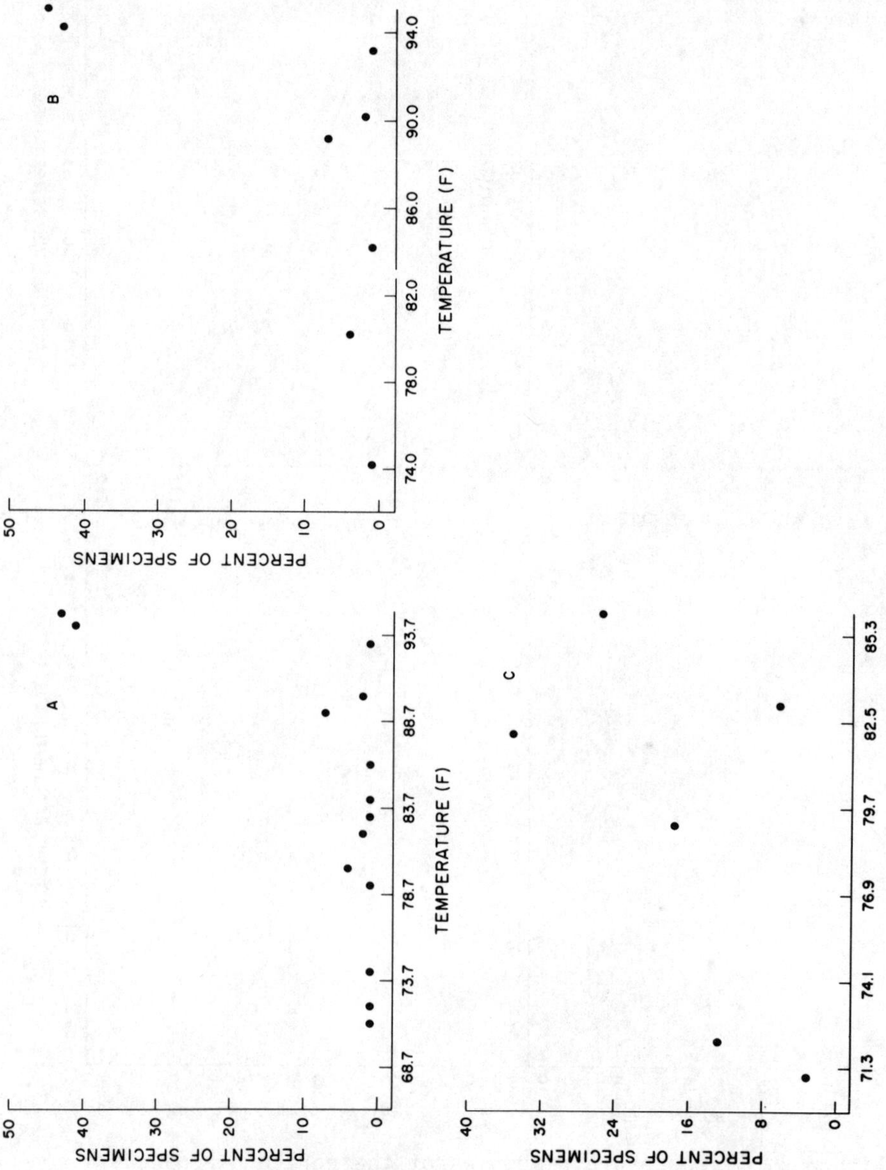

Figure 6. Relative abundance (corrected for sampling effort) of the channel catfish (<u>Ictalurus punctatus</u>, Rafinesque); A - composite of 1973 and 1974 data; B - 1973 data; C - 1974 data.

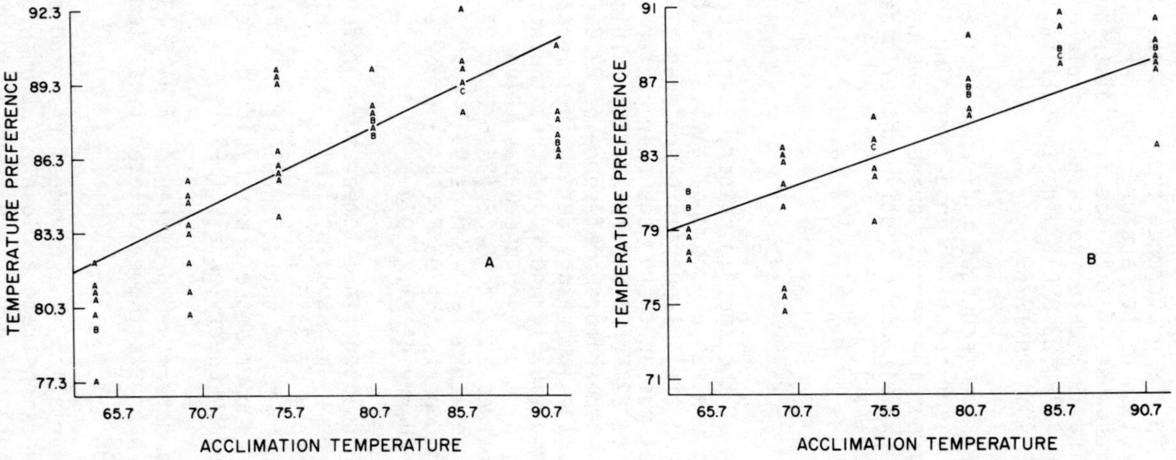

Figure 7. Laboratory relationship between acclimation temperature and temperature preference for: A - smallmouth bass (Micropterus dolomieui, Lacepede) and B - spotted bass (Micropterus punctulatus, Rafinesque), where A = 1 observation, B = 2 observations, etc.

The fantail darter (Etheostoma flabellare, Rafinesque) is an example of a species which did not respond in the laboratory experiments. When placed in the preference trough it would move to the nearest corner regardless of temperature. However, the field data showed a definite response of this species' distribution to temperature (Fig. 8). The 1973 data indicated a preference for temperatures cooler than 75°F (23.8°C). The fantail darter was present only at 67°F (19.4°C) in 1974.

The stepwise regression technique indicated that temperature was the single most important parameter in determining the distribution of all species except the spotfin shiner, none of the variables tested adequately explained the spotfin shiner's distribution (Stauffer et al., with editor).

DISCUSSION

With the exception of the spotfin shiner, which appears to be extremely euryoecious, there was a good qualitative relationship between the laboratory and field temperature preference data. The field data indicated that the northern hogsucker preferred water temperatures below 80°F to 82°F (26.6°C to 27.7°C). This agrees very well with the final temperature preference of 82.2°F (27.8°C) which was calculated from the laboratory data. These data also support the work of Gammon (1973), who found that the northern hogsucker avoided a thermal discharge on the Wabash River. Although the preferred temperature was 80°F to 82°F (26.6°C to 27.7°C) it was present in water temperatures as warm as 95°F (35°C) under field conditions, which supports the 95°F (35°C) upper avoidance temperature as found by Cherry et al. (1974).

The 80°F (26°C) calculated final temperature preference of the stoneroller also agrees reasonably well with the field determined preference temperature of 73°F to 75°F (22.7°C to 23.8°C). The presence of some specimens of the stoneroller in 93°F (33.8°C) water temperatures again supports the upper avoidance temperature of 91°F (32.7°C) reported by Cherry et al. (1974).

Finally, the field determined preference of 94°F (34.4°C) for the channel catfish supports the calculated final temperature preferendum of 92.8°F (33.7°C).

Despite inherent variation of individual fish and errors in sampling, the above data demonstrate a close relationship between field determined and laboratory determined temperature preference. Because of this relationship, it seems reasonable to make generalizations about the temperature responses of species for which only

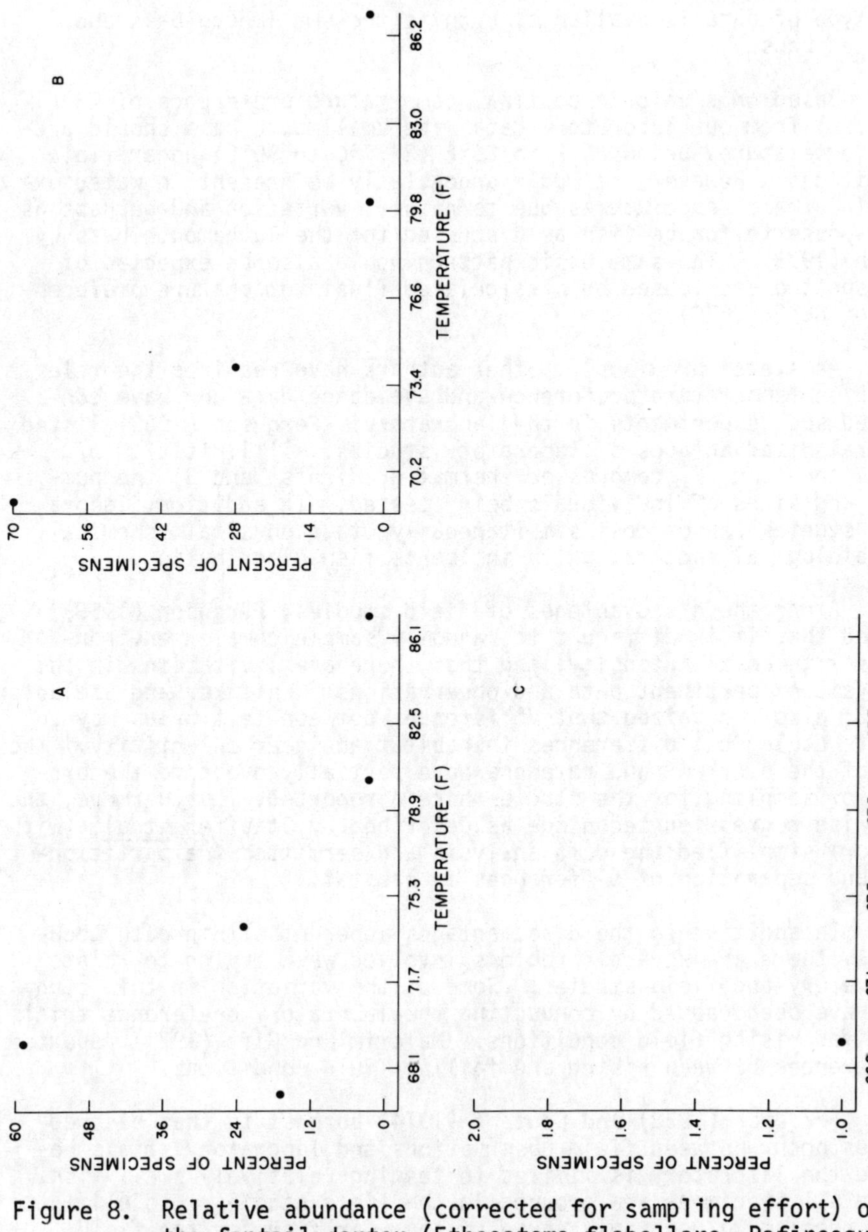

Figure 8. Relative abundance (corrected for sampling effort) of the fantail darter (Etheostoma flabellare, Rafinesque); A - composite of 1973 and 1974 data; B - 1973 data; C - 1974 data.

one type of data is available, such as the smallmouth bass and spotted bass.

Based on a calculated final temperature preference of 83°F (38.3°C) from our laboratory data, the smallmouth bass should prefer temperatures below 82°F to 86°F (27.7°C to 30°C) under field conditions. However, it would undoubtedly be present in water exceeding these temperatures due to natural variation and perhaps as a response to forage fish as discussed for the largemouth bass by Hatch (1973). The same basic pattern would also be expected of the spotted bass based on a calculated final temperature preferendum of 86°F (30°C).

As stated previously, other authors have realized the relevance of temperature preference and avoidance data and have conducted such experiments in the laboratory. Ferguson (1958) listed several disadvantages of laboratory studies: 1) limitation of space and time, 2) compressed thermal gradients, and 3) the numbers and sizes of individuals being tested. In addition, laboratory studies cannot test simultaneously other physical, chemical, and biological factors, which influence fish distribution.

Among the disadvantages of field studies, Ferguson (1958) stated that it is difficult to randomly sample complex environments with equal intensity, and that there are limitations in the analysis of pertinent data and observations. Whitaker and Schlueter (1973) also recognized that differences between test areas may be due to ecological differences in habitat and gear selectivity. The use of the blocknet and rotenone have partially overcome the problem of sampling for the studies herein reported. Furthermore, the stepwise regression technique as described by Stauffer et al. (with editor) simplified the data analysis and permitted the partitioning and separation of differences in habitat.

In addition to the disadvantages inherent within each technique, there are several problems involved when trying to relate laboratory and field studies. Some of the variation in this study may have been caused by conducting the laboratory preference trials only for rising field conditions. Meldrim and Gift (1971) reported differences between rising and falling field conditions.

Ferguson (1958) and Coutant (1974) both state that discrepancies occur between field observations and laboratory trials because the laboratory is limited to testing relatively small fish, while all specimens are grouped in the field studies. To eliminate this problem we used only young of the year fish for the field determinations of the channel catfish and northern hogsucker. However, one, two, and three year old specimens of the spotfin shiner and the stoneroller were used, since all age classes could be tested easily in both the laboratory and the field.

Another major advantage of the field studies is that in addition to determining temperature preferendum, they can provide information about population structures, year class differences, community structure changes, and relative growth rates without further sampling effort.

The on-site laboratory studies representing part of this investigation also provided specific advantages: 1) pumping river water directly into the laboratory minimized water quality changes between laboratory and field conditions, 2) handling time and transportation distances for fish were greatly reduced, 3) duplication of dynamic field conditions was attained more easily under the semi-controlled laboratory situation.

Through the use of both laboratory and field determined temperature responses, one could describe adequately the temperature relationships of the important fish species for this area of the New River. Fish responded in five ways: 1) some avoided warm temperatures and had a relatively cool ultimate temperature preferendum (i.e., 80°F to 85°F or 26.6°C to 27.7°C), such as the northern hogsucker and the stoneroller, 2) some were attracted to warm temperatures, and had a relatively warm ultimate preference (93.8°F or 34.3°C), such as the channel catfish, 3) some showed a preference temperature-acclimation temperature relationship in the laboratory, but demonstrated no preference temperature in the field, such as the spotfin shiner; 4) some, such as the smallmouth bass and the spotted bass, which had a distribution potentially affected by forage fish distribution (Hatch, 1973) and which were not present in sufficient abundance to make a field determination, but did respond to laboratory conditions, and 5) some, such as the fantail darter, which did not demonstrate any response under laboratory conditions, but did have a distribution which was markedly controlled by temperature under field conditions.

Temperature selection data similar to the above should be considered in the establishment of thermal standards on a site specific basis. This is especially true for situations similar to the New River, where the total heated discharge is small compared to the average annual flow of the river. This approach would be of special interest in warm water fisheries if it could be demonstrated that: a) refuge areas are available, b) important species, as defined by the EPA (1973), prefer the warm water (such as the channel catfish in this study) or will avoid the warm water during periods when the effluent temperatures may exceed their ultimate upper incipient lethal temperature, maximum temperature for optimum growth, maximum temperature for optimum reproduction, or maximum temperatures for short term exposures, c) the avoidance of thermally influenced areas by important species does not significantly decrease the area needed for successful reproduction, growth and maintenance of a viable population.

Basically, the use of site specific laboratory and field data make it possible to understand the effects of heated discharges. Information on preference and avoidance and population structure when combined with physical information under a variety of flow and operational patterns, make a determination of the impact on the aquatic community possible. This this information available, rational decisions can then be made about what kinds of thermal limitations are justifiable and appropriate.

We are not proposing that temperature preference data replace the existing criteria, but that they be considered along with other criteria. Based on our laboratory and field temperature preference studies at Appalachian Power Company's Glen Lyn Plant, no appreciable harm to the fish community should occur even though the thermal discharges are approximately 14°F (7.7°C) above ambient river temperatures and reach a maximum of 95°F in the plume. Therefore it seems evident that setting of an upper stream temperature standard of 81°F (27.2°C) is not scientifically justifiable under present conditions on the New River. This decision is also supported by Stauffer et al. (in press), who studied fish community structure, population structure, and condition factors at the same site.

It appears that the fish community can maintain its structure and function with the presence of some thermal discharges, provided adequate areas of ambient temperature are available. Thus our data support the approach contained in the 316(a) variance procedures for alternatives to closed cycle cooling (Federal Register, March 28, 1974) which allows an on-site demonstration of thermal effects to take precedent over an arbitrarily established stream standard.

CONCLUSIONS

1. There was a relationship between laboratory and field determined temperature preference data for the stoneroller, northern hogsucker, and channel catfish.

2. On the basis of the above relationship, one could predict the temperature responses in situ of the smallmouth bass and spotted bass from laboratory data.

3. Temperature preference data of important species indicated that a thermal discharge of 528 cfs (14.95 m^3/sec) which was 14°F (7.7°C) higher than ambient temperatures would produce no appreciable harm to the maintenance of a viable fish community. This conclusion was supported by the work of Stauffer et al. (in press) on the same discharge.

4. Temperature preference and avoidance data coupled with the EPA's (1973) proposed criteria can provide valuable insight toward the establishment of standards on a site specific basis.

LITERATURE CITED

BERKOWITZ, D. A. and A. M. SQUIRES. 1969. Symposium of the Committee on Environmental Alteration, American Association for the Advancement of Science. In Power Generation and Environmental Change. 440 pp.

BREEDER, C. M. and D. E. ROSEN. 1966. Modes of Reproduction in Fishes. The Natural History Press. Garden City, N.Y. 941 pp.

BRETT, J. R. 1952. Temperature tolerance in Young Pacific Salmon, genus Oncorhynchus. J. Fish. REs. Bd. Can. 9: 265-323.

BRYSON, W. T., R. T. LACKEY, J. CAIRNS, JR. and K. L. DICKSON. In press. Restocking after fish kills as a fisheries management strategy. Trans. Am. Fish. Soc.

CAIRNS, J. 1972. Coping with heated waste discharges from steam-electric power plants. Bio-Science. 22(7): 411-420.

CHERRY, D. S., K. L. DICKSON and J. CAIRNS, JR. 1974. The use of a mobile laboratory to study temperature response of fish. 29th Purdue Industrial Conference. 28 pp.

COUTANT, C. C. 1970. Biological aspects of thermal pollution. I. Entrainment and discharge canal effects. CRC Critical Reviews in Environmental Control 1(3): 341-381.

COUTANT, C. C. 1972. Biological aspects of thermal pollution. II. Scientific basis for water temperature standards at power plants. CRC Critical Reviews in Environmental Control 3(1): 1-24.

COUTANT, C. C. 1974. Temperature selection by fish--A factor in power plant impact assessments. Symposium on the Physical and Biological Effects on the Environment of Cooling Systems and Thermal Discharges at Nuclear Power Stations. 25 pp.

DENDY, J. S. 1948. Predicting depth distribution of fish in three TVA storage type reservoirs. Trans. Am. Fish. Soc. 75: 88-142.

DOUDOROFF, P. 1938. Reactions of marine fishes to temperature gradients. Biol. Bull. 75: 494-509.

ENVIRONMENTAL PROTECTION AGENCY. 1973. Proposed criteria for water quality. 1: 144-170.

FEDERAL REGISTER. 1974. Thermal discharges 40 CFR Part 122, 39 FR, 11434 (March 28).

FERGUSON, R. G. 1958. The preferred temperature of fish and their midsummer distribution in temperature lakes and streams. J. Fish. Res. Bd. Can. 15(4): 607-624.

FISHER, K. C. and P. ELSON. 1950. The selected temperature of Atlantic salmon and speckled trout and the effect of temperature on the response to an electrical stimulus. Physiol. Zool., 23: 28-34.

FRY, F. E. J. 1947. Effects of the environment on animal activity. Univ. Toronto Stud., Biol. Serv., No. 55. 62 pp.

GAMMON, J. R. 1973. The response of fish populations in the Wabash River to heated effluents. 3rd Symposium on Radioecology. D. J. Nelson ed.

GARSIDE, E. T. and J. S. TAIT. 1958. Preferred temperature of rainbow trout (Salmo gairdneri, Richardson) and its unusual relationship to acclimation temperature. Can. J. Zool. 36: 563-567.

HATCH, J. T. 1973. The responses of fish fauna of Little Three Mile Creek, and the Ohio River to a thermal effluent. Master's Thesis. De Pauw Univ. 91 pp.

HOCUTT, C. H., P. S. HAMBRICK, and M. T. MASNIK. 1973. Rotenone methods in a large river system. Arch. Hydrobiol. 72(2): 245-262.

KENNEDY, W. A. 1941. The migration of fish from a shallow lake to a deep lake in spring and early summer. Trans. Am. Fish. Soc. 70: 391-396.

MELDRIM, J. W. and J. J. GIFT. 1971. Temperature preference, avoidance and shock experiments with estuarine fishes. Ichth. Assoc. Bull. 7. 75 pp.

PARKER, F. L. and P. A. KRENKEL. 1969. Engineering Aspects of Thermal Pollution. Vanderbilt Univ. Press, Nashville, Tenn. 351 pp.

RANEY, E. C. and B. W. MENZEL. 1969. Heated effluents on aquatic life with emphasis on fishes. Ichth. Assoc. Bull. 2. 470 pp.

RANEY, E. C., B. W. MENZEL, and E. C. WELLER. 1973. Heated effluents on aquatic life with emphasis on fishes, a bibliography. Ichth. Assoc. Bull. 9. 651 pp.

SOKAL, R. R. and F. J. ROHLF. 1969. Biometry. W. H. Freeman and Company, San Francisco. 776 pp.

STAUFFER, J. R., K. L. DICKSON, J. CAIRNS, JR., W. F. CALHOUN, M. T. MASNIK and R. H. MYERS. (With editors). The determination of fish temperature preference and avoidance from field data via stepwise linear regression techniques.

STAUFFER, J. R., K. L. DICKSON and J. CAIRNS, JR. In press. A field evaluation of the effects of heated discharges of fish distribution. Water Res. Bull.

WHITAKER, J. O., JR. and R. A. SCHLUETER. 1973. Effects of heated discharge on fish and invertebrates of White River at Petersburg, Indiana. Water Resources Research Center. Report No. 6. 123 pp.

EXPERIMENTS RELATED TO DIRECTING ATLANTIC SALMON SMOLTS, *SALMO SALAR*, AROUND HYDROELECTRIC TURBINES

J. Richard Semple and Curtis L. McLeod
Resource Development Branch, Fisheries and Marine Service
Department of the Environment, Halifax, Nova Scotia

ABSTRACT

At a hydroelectric dam on Tusket River, Nova Scotia, a fish bypass was constructed through the powerhouse and a vacant turbine seating so that fish would be able to avoid the turbines. An evaluation of this facility and a floating-screen deflector, using hatchery-reared Atlantic salmon smolts, *Salmo salar*, revealed that bypass efficiency was significantly associated with the time during the smolt migration season at which the experiments were conducted, $P<0.001$. Late season experiments gave larger numbers of smolt recoveries in the bypass. Guidance efficiency was also dependent on the deflector angle ($P<0.001$) and the depth of screens used ($P<0.001$) It increased when screening depth was increased, and the deflector angle was reduced in relation to the direction of river flow. The best deflector array tested operated to a depth of 2.44 m. without a sounding lip, and at an angle of 45°. This arrangement gave an average bypass efficiency of 72.1 percent and was the only deflector array to supersede the efficiencies of late season control tests by more than 15 percent. Average bypass efficiencies of 1973 and 1974 control tests were 18.5 percent and 48.2 percent, respectively.

Significant correlations existed between turbine discharge ($P<0.01$), bypass discharge ($P<0.1$), and smolt passage rate (smolt/hr) through the bypass. The floating-screen deflector modified these relationships so that no significant correlations were found at similar probability levels. Tests indicated that smolts best utilized the bypass when entrance depth exceeded 19 cm. and average flows of 0.32-0.53 m^3/sec. were maintained.

INTRODUCTION

The purpose of this study was to determine the usefulness of a fish bypass and floating-screen deflector for directing Atlantic salmon smolts around hydroelectric turbines at a power station near the mouth of Tusket River in southwestern Nova Scotia. It was known in advance that smolt-sized juvenile salmon suffered losses of 16.5 percent[1] on passing through turbines at this hydroelectric facility.

The river has a total length, including significant tributaries, of 644 km. and a drainage area of 1,450 km. Over 130 lakes are contained within the river system. In its present state, the river is estimated to have a salmon production capability of 1,040 to 2,080 adults annually.[2]

REVIEW OF PAST EXPERIENCE

Recent reviews concerning provisions for safe fish passage around hydroelectric dams and other potentially hazardous migration routes include those by Eicher (1970) and Burns (1966). The work by Clay (1961) comprehensively summarizes all fish passage technology up to the time it was written. Huber (1974) and Scharma (1973) have compiled pertinent bibliographies of fish protection at water diversions, intakes and dams.

In the present context, fish directing and bypass systems all have the same objective--the protection of fish. Many methods have been attempted, including light, sound, bubbles, dangling chains, mechanical screens, electrical screens, and louvers for guiding or preventing downstream-migrant fish from entering hazardous areas and for directing them into a safe bypass. The latter three appear to offer the best prospects for accomplishing this end under varying flows and accompanying conditions. Only these are described in the following brief review.

Mechanical Screens

Rotary drum screens have met with varying success. Fish may pass over or through the screens when water velocities exceed 0.46 m/sec. (Campbell, 1959) or the depth of submergence increases (Clay, 1961). Ordinarily screens of this type are limited to a

---

[1]Smith, K.E.H. 1960. Mortality tests on young salmon and gaspereau, Tusket River Dam, Yarmouth County, Nova Scotia. Can. Dept. Fish. Unpubl. Rept. 37 pp.

[2]Anon. 1962. Tusket River survey, Yarmouth County, Nova Scotia. Can. Dept. Fish. Unpubl. Rept. 37 pp.

depth of 4.6 m. (Wales et al., 1950). Fields (1957) reports that 50 to 70 percent of juvenile chinook and silver salmon were carried over drum screens on the White River diversion in Washington. Campbell (1959) obtained better results with similar screens in northeastern Oregon, where he reported efficiencies of 14 to 82 percent in one series of tests with juvenile steelhead trout, and 20 to 100 percent efficiency (usually greater than 88 percent) in another series of experiments with juvenile chinook salmon and trout. In California, installation of rotary drum screens has ceased due to high fish escapement under or over the screens and serious maintenance problems (Burns, 1966).

Many other types of mechanical screens, which are more applicable to small streams or river diversions, have been described in the literature (Kupka, 1966; Leitritz, 1952; Wales et al., 1950; Whalls et al., 1957) without indicating their fish exclusion or guidance efficiency.

One of the most promising developments in high efficiency fish screening is the horizontal traveling screen which was conceived and tested by the United States Bureau of Commercial Fisheries (Bates, 1970). Application of this screen in the Umatilla River irrigation canal in Oregon resulted in fish deflection efficiencies of 97 to 100 percent. Features which make this screen so attractive are its high efficiency under different water levels and water velocities, and its reduction of injury to fish which may become impinged on the screens.

At large run-of-the-river dams, fish skimmers and gulpers (artificial outlets) have been used with some success. In these devices, a flow of 1.4-11.3 m^3/sec. is drawn into the facility and passed through a horizontal inclined screen or a vertical traveling screen set at an angle to the flow. Most of the flow is pumped back into the headpond for power generation purposes (Eicher, 1970). A small amount of water, usually about 0.14 m^3/sec., passes over a weir on the downstream end of the facility along with any screened fish. The fish are transported to the tailrace by different methods.

Skimmers differ from gulpers in being fixed at one location. Gulpers are floating structures that can be moved to different areas to intercept fish. Fish interception efficiencies for skimmers have ranged from a few percent to 60 percent on the Clackamus River (Eicher, 1970). The efficiencies of gulpers operated in reservoirs on the Baker, Lewis and Cowlitz rivers in Washington have ranged from satisfactory in the former, to poor to mediocre in the latter two rivers.

Screens have been used at the entrance to turbine intakes. At hydroelectric installations in Scotland it has become common practice to screen turbine intakes of dams with over 30.5 m. head (Pyefinch, 1966). The benefits of such a measure are dubious since the overall reaction of fish is not to seek another outlet but to be held in the impoundment.

Scale-model simulation studies have also been conducted on the usefulness of screens for guiding fish into gatewell entrances in turbine intakes (vertical chambers rising from the ceiling of an intake to the forebay level and used to lower gates and prevent water reaching the turbines). High-porosity gatewell screens operated at an angle of 45° to the flow, and adjusted in length to screen two-thirds of the discharge, have been up to 87 percent efficient (Marquette and Long, 1971) in diverting juvenile chinook salmon. Before this method of screening becomes practical, it must be proven that the screens can remain safely for long periods of time without cleaning, and without washing out and damaging the turbines (Long and Krema, 1969). Simulation results must also be confirmed at a full-sized hydroelectric installation.

Attempts have been made to divert and collect fish with long leads (Pugh et al., 1969) or partial screens. One such experiment, conducted in the Brownlee Reservoir on the Snake River, was unsuccessful because fish sounded under the 6.1-9.1 m. deep netting. Another experiment at Baker dam employed a 73.2 m. long by 4.6 m. deep screen to guide juvenile salmon into a bypass trap. It increased the numbers of juvenile coho salmon entering the bypass (Ress, 1957; original paper not seen) significantly.

To date, fish protection provisions incorporating mechanical screens are most widely used and generally with better success (Eicher, 1970).

Louvers

Louver diverters consist of a fence-like series of vertical, steel slats. The fence is usually installed at an angle of 12 to 20° to the direction of flow and leads to a fish bypass. Water strikes the slats perpendicularly. Fish approaching the louvers tend to avoid them and move into the bypass.

The fish guiding efficiency of louvers is generally related to slat spacing (Bates and Jewett, 1971; Monan et al., 1967), approach/bypass velocity ratios (Ducharme, 1972; Ruggles and Ryan, 1964), approach velocity (Bates and Jewett, 1961; Ducharme, 1972), and the species and age of fish dealt with (Ruggles and Ryan, 1964).

Wider slat spacing (5.1 to 30.5 cm.) did not impair guiding efficiency for Atlantic salmon smolts (Ducharme, 1972).

Bates and Jewett (1961) report guiding efficiencies of 62.6 to 98.6 percent for juvenile steelhead trout, and Monan et al. (1967) have recorded efficiencies of 56.3 to 84.4 percent for the same species, using non-electrified louvers. Ruggles and Ryan (1964) indicate that one can expect louver guiding efficiencies of 85 to 95 percent for sockeye and coho salmon smolts, 65 to 75 percent for chinook fry, and 20 to 30 percent for coho smolts using a single line louver array under the conditions tested. Ducharme (1972) was able to increase the average efficiency of louvers for guiding Atlantic salmon smolts, from 57 percent to 80 percent by modifications which increased velocity and reduced turbulence in the bypass. He concluded that louvers are a practical method for guiding Atlantic salmon smolts from water flowing at 0.61 m/sec. or more in a narrow channel.

Electrical Screens

Electrical screens have been successful in diverting downstream migrant fish only under certain conditions. For example, a small electrical screening installation on Swelzer Creek, British Columbia was over 90 percent efficient for juvenile sockeye salmon. When similar screens were employed at Baker Dam in Washington, no guiding success was achieved for either juvenile sockeye or coho salmon (Andrew et al., 1955).

Other tests revealed the necessity of low flows, with velocities less than 0.30 m/sec. to obtain good electrical guiding efficiency (Pugh et al., 1969). When these conditions were met, screen efficiency ranged from 70 to 80 percent, but at higher flows 40 to 55 percent efficiency usually resulted. These authors concluded that the electrical screen would not be practical for diverting juvenile steelhead trout or coho and chinook salmon in most rivers and streams.

In Great Britain, Lethlean (1953) obtained guiding efficiencies of up to 70 percent for Atlantic salmon smolts, but conceded that more testing is required to make electrical screens a practical option for guiding Atlantic salmon smolts.

DESCRIPTION OF HYDROELECTRIC INSTALLATION

The Tusket River power installation was constructed in 1929. It consists of a powerhouse dam located on an impounded portion of a 30 to 53 m. wide, unlined, diversion channel. A diversion dam is located about 0.9 km. upstream from the powerhouse on the original watercourse. The flows from both dams converge 0.6 km. downstream from the powerhouse. Pool-and-weir fishways exist at both dams (Fig. 1).

The power plant is equipped with three Kaplan turbines capable of producing 900 KVA when operated at a maximum head of 8.2 m. Except for sporadic spillway discharge during May and June, most of the river flow is used for power generation. During these months, all generating units are usually operated on a 24-hour basis.

FACILITIES AND CONDITIONS DURING TESTING

Downstream fish passage facilities at the Tusket River hydroelectric installation were provided by constructing an entrance through the powerhouse forebay wall (Fig. 2), routing fish through two consecutive pools, and discharging them into the tailrace through a vacant draft tube opening. Because of tidal fluctuations, the head drop into the tailrace varied from 1.4 to 3.3 m. The downstream bypass was designed to accommodate a water flow of approximately 0.57 m^3/sec.

The bypass entrance (Fig. 2 - detail) is a funnelled opening, 2.3 m. wide on the upstream side and tapering to 1.6 m. at the discharge point. With 40 cm. depth at the entrance, water accelerates from nearly 70 cm./sec. at the mouth to 145 cm./sec. as it drops into the first pool.

Trapping facilities (Fig. 3) for the study consisted of a V-shaped inclined trough extending between pools 1 and 2 of the bypass. This system strained all water and fish passing into the second pool. Fish slid down the inclined trough onto a sorting rack and into a fiberglass holding tank. The sorting rack was found necessary to exclude large numbers of spent alewives (*Alosa pseudoharengus*) and permit continuous trapping of salmon smolts. The rack proved to be over 98 percent efficient in sorting smolts from larger fish.

In order to guide downstream-migrant fish away from the turbine intakes and into the bypass, an experimental floating-screen deflector was extended upstream from the powerhouse forebay wall. The deflector consisted of a series of individual polystyrene billets, decked with plywood to make a 1.83 m. by 2.44 m. platform. I-beam guides were attached vertically to the upstream side of each platform unit. Screen panels, 1.22 m. wide by 2.44 m. long were

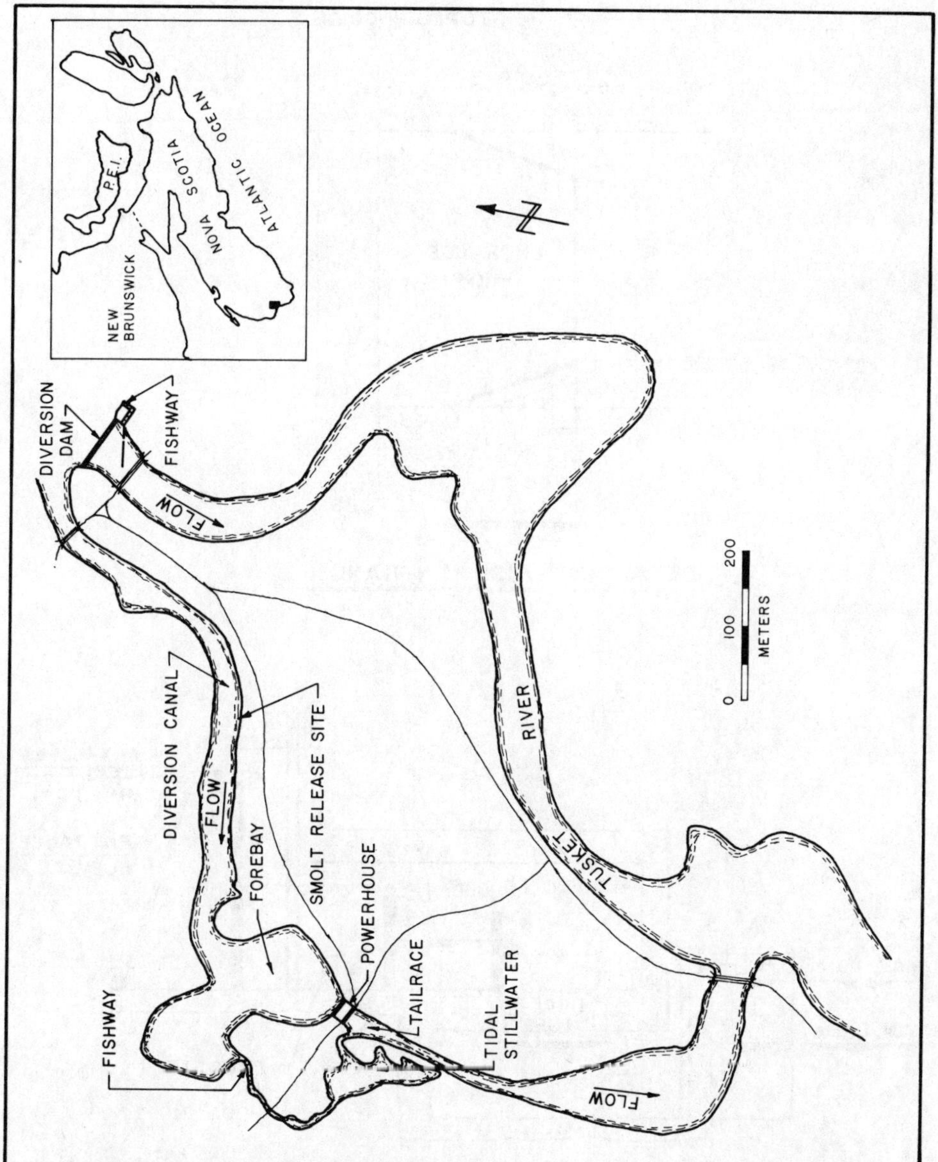

Figure 1. Geographic location and site plan of Tusket River hydroelectric facilities.

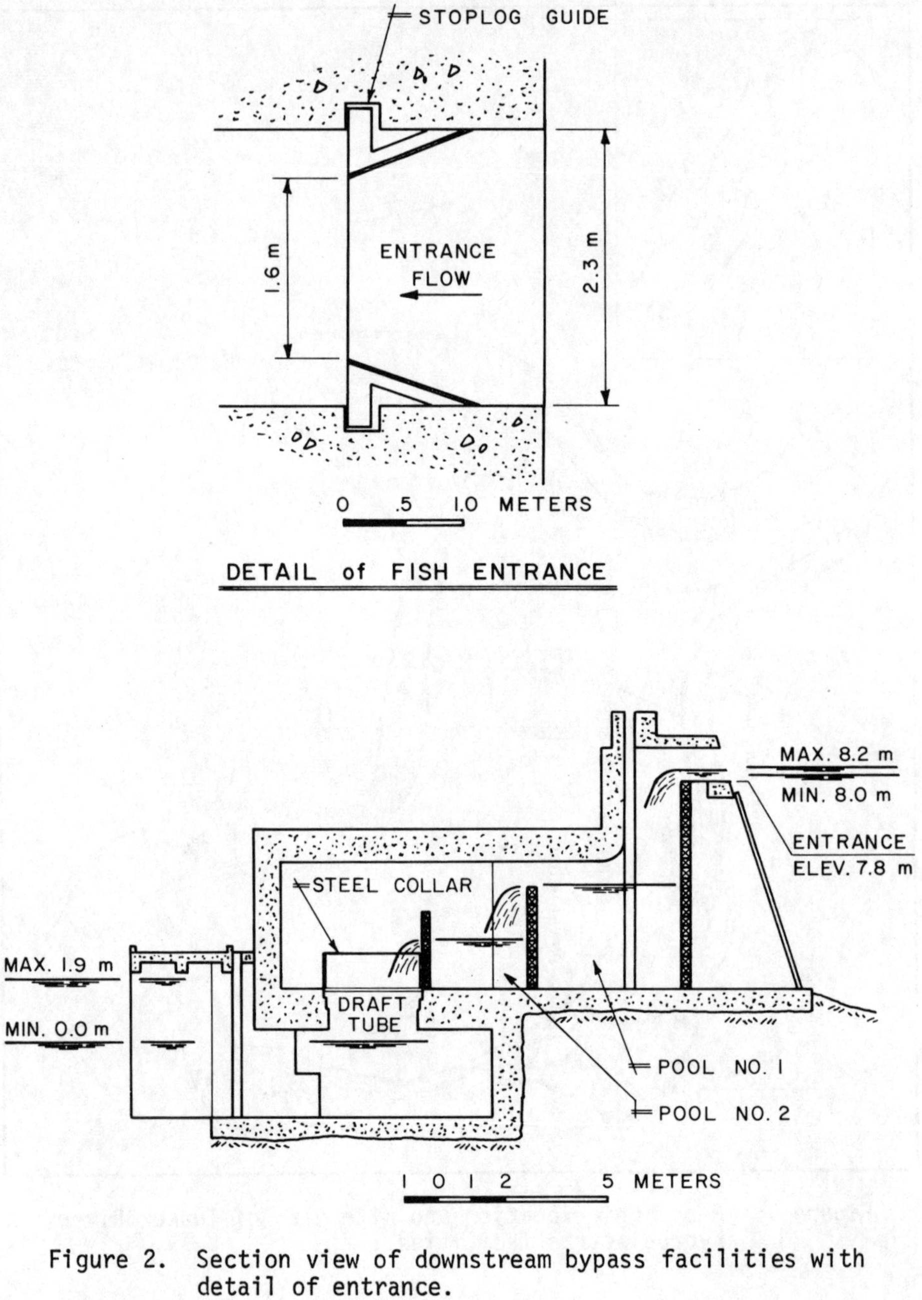

Figure 2. Section view of downstream bypass facilities with detail of entrance.

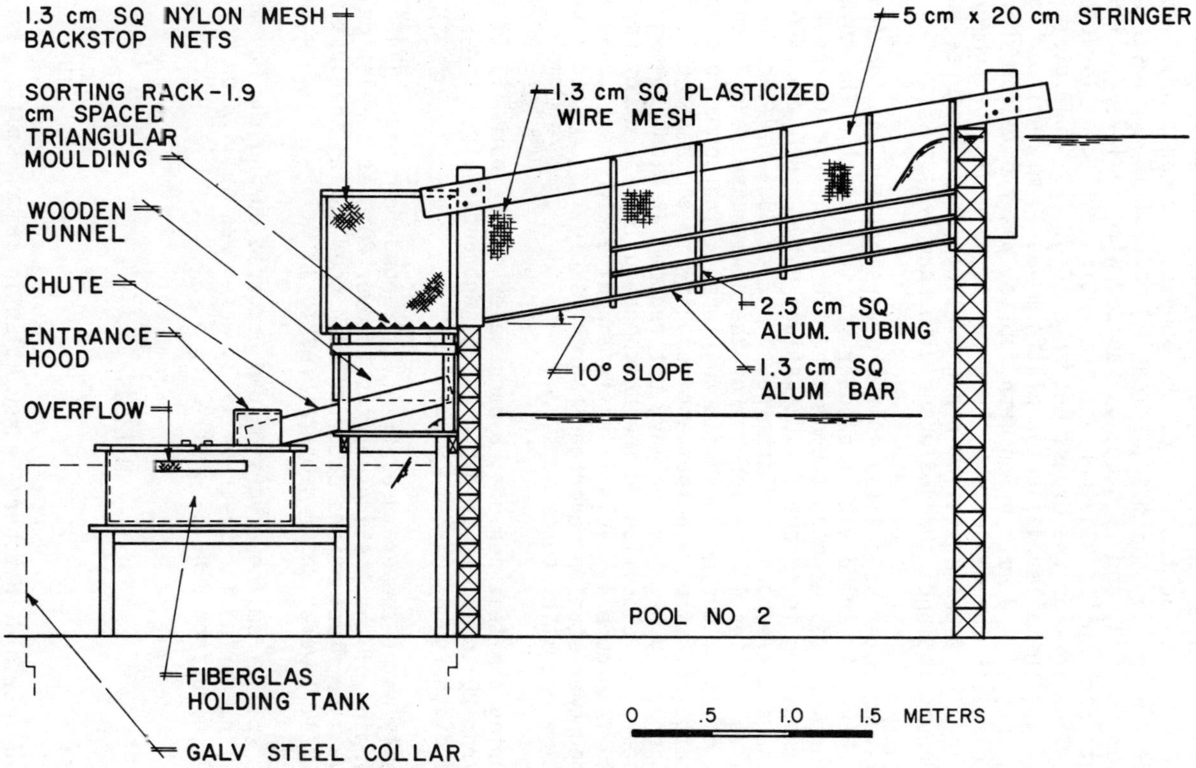

Figure 3. Detail of downstream bypass trapping facilities.

inserted in the guides and could be raised or lowered to the desired depth. In 1974 a 45 cm. wide sounding lip was installed at a right angle to the bottom edge of the panels. Screen panels were constructed with 1.3 cm.^{2} mesh, screening made of 18 kg. test knotless nylon in 1973, and 19 gauge galvanized wire in 1974. All platform units were bolted together and the entire deflector located to guide fish towards the bypass entrance (Fig. 4). The total length of the installed deflector platform was 40 m. A 1.8 m. deep floating net, of 1.3 cm.^{2} nylon-mesh fabric, was run to the shoreline to complete the deflector. In order to maintain the submerged screens in a vertical position in high velocity areas, it was necessary to counterbalance the deflector with sand bags.

Average water velocities in the diversion canal and powerhouse forebay (Fig. 4) varied with the power station output. The turbine generating load was dependent upon the river discharge and power demand, resulting in frequent fluctuations between 600 and 2400 KW output and a corresponding discharge of 9.2 to 35.4 m.^{3}/sec. Approach velocities in the forebay of the powerhouse ranged from 17 to 55 cm./sec. during these fluctuations. With normal precipitation patterns, turbine output and velocity in the powerhouse forebay was generally high during May, but reduced at night during June. Average daytime velocities in the headpond during the study period are shown in Figure 4. Velocities in the diversion canal, 450 m. above the power station, averaged approximately 130 cm./sec. during the daytime operating conditions.

During the test periods, bypass water depths normally fluctuated from 20 to 40 cm. in depth at the entrance weir, corresponding to a discharge of 0.23 to 0.65 m.^{3}/sec.

Daily water temperatures (Fig. 5) ranged from nearly 14°C at the beginning of the testing periods to over 20°C at completion. The only major fluctuations occurred when snow during the last week of May 1974, lowered water temperatures to 11°C.

No spill from the diversion dam occurred during the May and June testing period in 1974. In 1973, water was spilled on May 17, 18 and 24, and from June 24 until July 6.

METHODS

Procedure

Two year old, hatchery-reared Atlantic salmon smolts were employed as test fish. The fish were tagged at a nearby hatchery and released the following day in order to reduce the stress of tagging

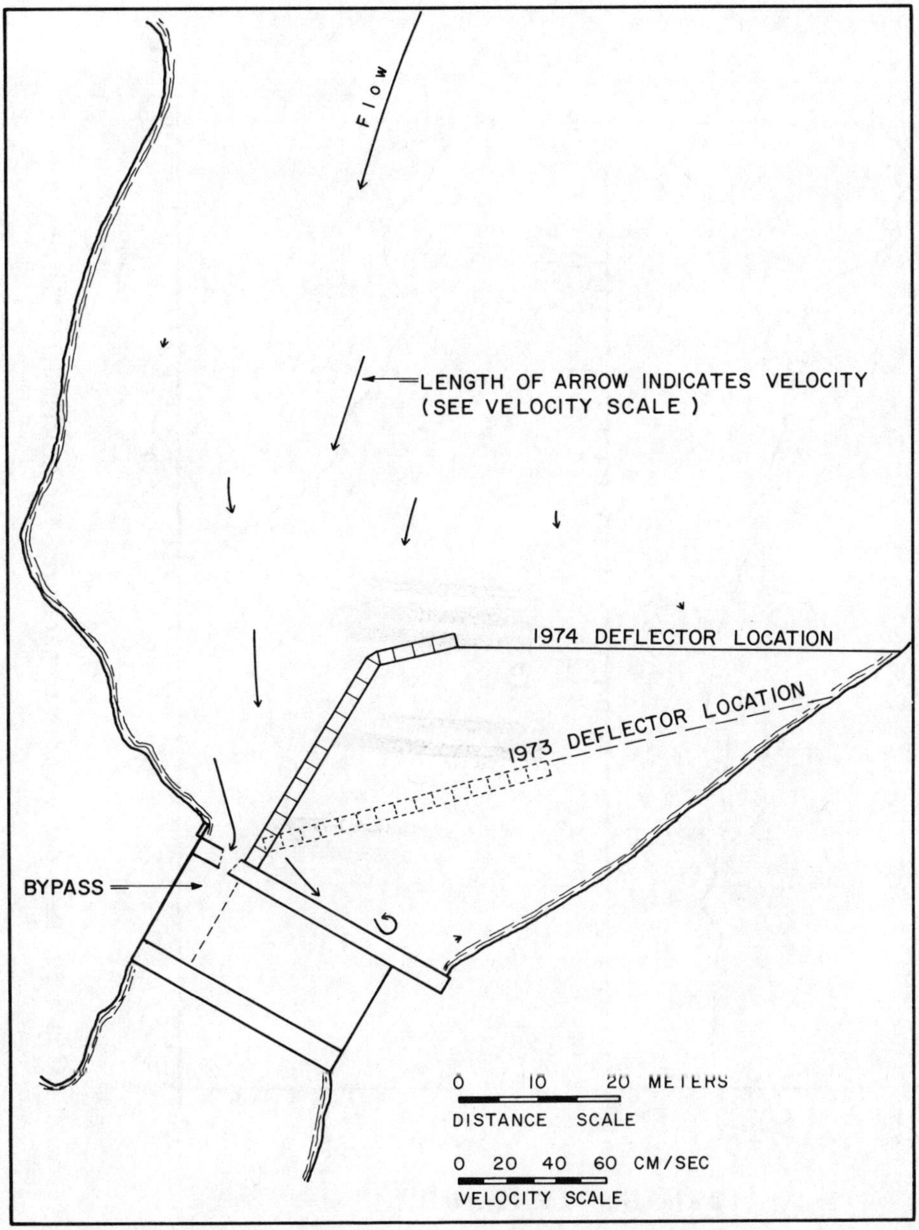

Figure 4. Deflector locations and mean headpond velocities during daylight hours.

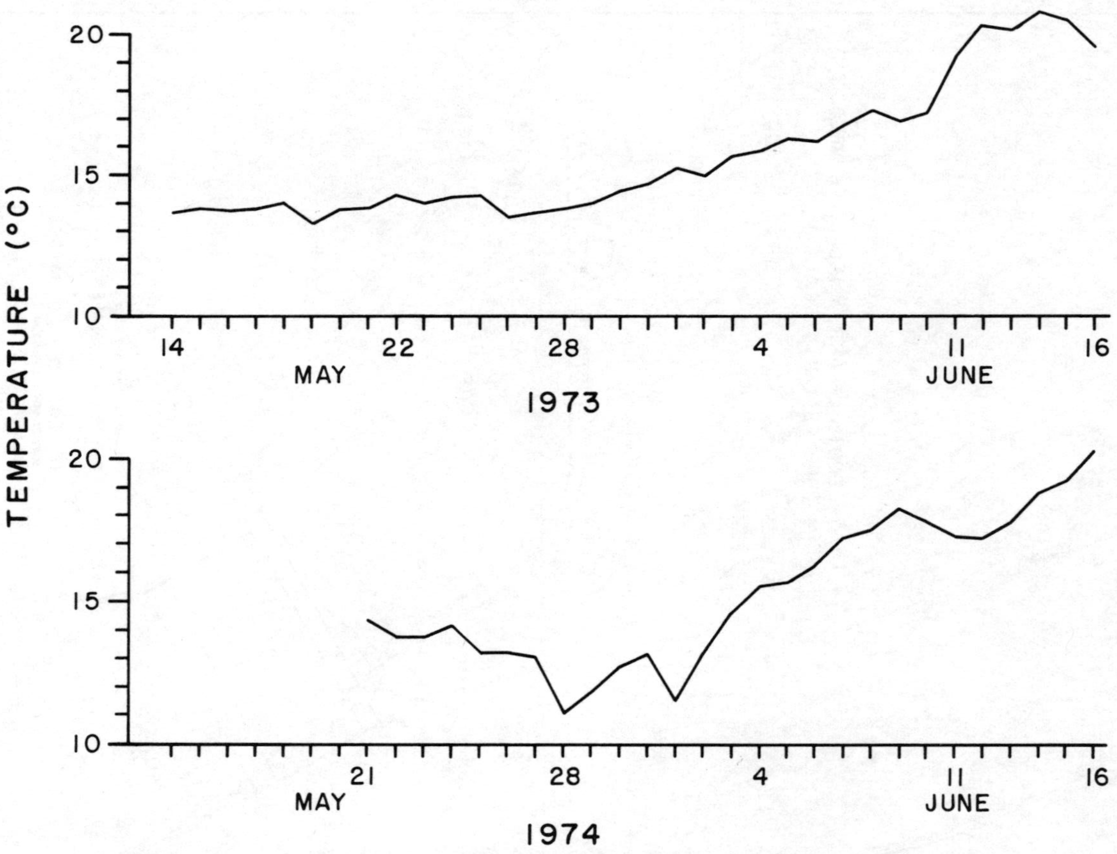

Figure 5. Mean daily water temperatures for headpond, 1973 and 1974.

and possible behavioral changes. Carlin tags were used to identify the fish, and were tied through the back, just anteriorly to the dorsal fin. The fish were trucked to the test site, and released in the diversion channel (Fig. 1).

A five day recovery period was employed to determine the efficiency of a test array for each lot of experimental fish. Three consecutive replications of each experiment were conducted in order to determine the variation in guidance efficiency of each test array. During the experiments a bypass trap was operated continuously but was usually only tended during daylight hours. Each test lot of fish was normally composed of 500 smolts in 1973 and 600 smolts in 1974.

Tests were conducted during May and June of both years to determine if smolt deflection efficiency could be improved by increasing screening depth, installing a sounding lip on the bottom edge of the screens or reducing the operating angle of the deflector in relation to the direction of flow. Control tests without deflectors were also implemented at the beginning and ending of the experiments.

Analysis

To minimize error in determining the efficiency of control tests (no deflector) and various deflector conditions, experimental fish were released in close proximity to the test array (Fig. 1) in an attempt to lessen predation by birds and eels (Anguilla rostrata). Only smolts longer than 15 cm. were used in the experiments, since it is known that the average length of wild Atlantic salmon smolts in Canadian rivers is 12.5 to 17.5 cm. (Dymond, 1963). In order to compare the merits of different test arrays statistically, it was necessary to compensate for delayed migration of smolts during the testing period. This was done by considering recovery rates of only those test lots which gave at least nine-tenths of the total recoveries within the five-day test periods. G-tests of independence or association (Sokal and Rohlf, 1969) were employed to determine the significance of the null hypothesis.

Turbine and bypass discharges were examined to determine their effect on smolt recovery rates (smolts/hr.) in the bypass. Control tests and deflector experiments were examined separately in order to exclude any effect the deflector system may have had on this relationship.

FACTORS AFFECTING BYPASS EFFICIENCY

Progression in Smolt Migration Season

Early control tests (without deflectors) in 1973, resulted in a low average bypass efficiency of 5.5 to 10.2 percent (Table 1). Similar control tests in 1974 had a much higher efficiency, i.e., 29.7 percent (Table 2). Late control experiments gave higher bypass efficiencies than the early controls in both years, 41.2 and 53.3 percent. Recovery rates from all early control experiments ranged from 5.5 to 39.7 percent and averaged 18.5 percent. Recovery rates for all late control tests ranged from 41.2 to 53.3 percent and averaged 48.2 percent. Bypass efficiency was significantly dependent ($P<0.001$) on the time during the smolt migration season at which experiments were conducted (Tables 1 and 3).

Different Deflector Arrays

Deflector arrays tested in 1973 resulted in average bypass efficiencies of 11.5 and 22.8 percent (Table 1). In both instances, the deflector was operated without a sounding lip on the screens, and at an angle of 90° to the direction of flow. The lowest efficiency was recorded for screens operating to a depth of 1.22 m. In a second series of tests the screens were operated to a depth of 2.44 m. A comparison of both series of experiments revealed that guidance efficiency was dependent ($P<0.001$) on the deflector conditions, but was not improved by the provision of a sounding lip (Tables 2 and 3).

When the deflector was operated at a similar depth (2.44 m.), without a sounding lip, but at different angles in relation to the direction of flow (90° and 45°), it was found that guidance efficiency increased from 22.8 percent (Table 1) to 72.1 percent (Table 2). Bypass efficiency was significantly ($P<0.001$) associated with the deflector angle, and increased when the angle of deflection was reduced (Table 3).

A comparison of the two 1973 deflector arrays with early and late control tests (without deflectors) revealed that the average efficiency of early control tests conducted during May 22-24, was independent ($P>0.05$) of the bypass conditions when shallower depth (1.22 m.) screens were used, but was dependent ($P<0.001$) when deeper (2.44 m.) screens were employed (Tables 1 and 3). Average bypass efficiency increased from 10.2 percent in the control tests to 22.8 percent in the deflector experiments in which deeper screens were used. An examination of the bypass efficiencies of late control tests conducted in 1973 with both deflector arrays, revealed

Table 1. Release and recovery statistics for tagged Atlantic salmon smolts released under different test conditions to determine the usefulness of a fish bypass and floating-screen deflector, 1973.

Test number	Smolt releases		Smolt recoveries[1]			Test conditions
	Date	Number	Number taken within 5-day test period	Total number[2] taken	% of total recoveries taken in 5-day test period	
1	May 14	500	26(5.2)	46(9.2)	56.5	No deflector
2	15	500	24(4.8)	52(10.4)	46.2	
3	16	500	33(6.6)	46(9.2)	71.7	
Average	14-16	1500	83(5.5)	144(9.6)	57.3	
4	22	500	59(11.8)	60(12.0)	98.3	No deflector
5	23	500	44(8.8)	47(9.4)	93.6	
6	24	500	50(10.0)	56(11.2)	89.3	
Average	22-24	1500	153(10.2)	163(10.9)	93.6	
7	28	500	45(9.0)	45(9.0)	100.0	Deflector without lip, 1.22-m-deep screens, 90° deflection angle
8	29	500	79(15.8)	79(15.8)	100.0	
9	30	500	48(9.6)	51(10.2)	94.1	
Average	28-30	1500	172(11.5)	175(11.7)	98.3	
10	June 4	500	36(7.2)	36(7.2)	100.0	Deflector without lip, 2.44-m-deep screens, 90° deflection angle
11	5	500	147(29.4)	148(29.6)	99.3	
12	6	500	159(31.8)	163(32.6)	97.5	
Average	4-6	1500	342(22.8)	347(23.1)	98.7	
13	11	500	264(52.8)	264(52.8)	100.0	No deflector
14	12	450	127(28.2)	127(25.4)	100.0	
Average	11-12	950	391(41.2)	391(41.2)	100.0	

[1]Figures in parentheses are the percentage recoveries from the total tagged smolts released.
[2]Smolt trapping for each experiment was continuous from the data of release up to and including June 17.

Table 2. Release and recovery statistics for tagged Atlantic salmon smolts released under different test conditions to determine the usefulness of a fish bypass and floating-screen deflector, 1974

Test number	Smolt releases		Smolt recoveries[1]			Test conditions
	Date	Number	Number taken within 5-day test period	Total number[2] taken	% of total recoveries taken in 5-day test period	
1	May 21	600	219(36.5)	265(44.2)	82.6	No deflector
2	22	600	241(40.2)	312(52.0)	77.2	
3	23	600	254(42.3)	333(55.5)	76.3	
Average	21-23	1800	714(39.7)	910(50.6)	78.5	
4	28	600	338(56.3)	359(59.8)	94.1	Deflector with lip, 2.44-m-deep screens 45° deflection angle
5	29	600	174(29.0)	187(31.2)	92.9	
6	30	600	308(51.3)	343(57.2)	89.7	
Average	28-30	1800	820(45.5)	889(49.4)	92.1	
7	June 4	600	384(64.0)	385(64.2)	99.7	Deflector without lip, 2.44-m-deep screens, 45° deflection angle
8	5	600	461(76.8)	463(77.2)	99.5	
9	6	600	453(75.5)	453(75.5)	100.0	
Average	4-6	1800	1298(72.1)	1301(72.3)	99.7	
10	11	600	246(41.0)	247(41.2)	99.5	No deflector
11	12	600	389(64.8)	390(65.0)	99.7	
12	13	452	246(54.4)	246(54.4)	100.0	
Average	11-13	1652	881(53.3)	883(53.4)	98.4	

[1]Figures in parentheses are the percentage recoveries from the total tagges smolts released.
[2]Smolt trapping for each experiment was continuous from the date of release up to and including June 18.

Table 3. Statistical significance of independence tests for various test conditions and physical arrangement of deflectors.

Statistical comparison of different test arrangements[1] Condition A	Condition B	Calculated G-value	Degrees of freedom	Significance
No deflector (May 22-24/73)	90°, 1.22m (May 28-30/73)	1.25	1	P>.05
No deflector (May 22-24/73)	90°, 2.44m (June 4-6/73)	88.30	1	P<.001
No deflector (June 11-12/73)	90°, 1.22m (May 28-30/73)	285.65	1	P<.001
No deflector (June 11-12/73)	90°, 2.44m (June 4-6/73)	92.18	1	P<.001
No deflector (June 11-13/74)	45°, 2.44m, lip (May 28-30/74)	20.85	1	P<.001
No deflector (June 11-13/74)	45°, 2.44m (June 4-6/74)	131.22	1	P<.001
90°, 1.22m (May 28-30/73)	90°, 2.44m (June 4-6/73)	20.61	1	P<.001
45°, 2.44m (June 4-6/74)	45°, 2.44m, lip (May 28-30/74)	265.78	1	P<.001
90°, 2.44m (June 4-6/73	45°, 2.44m (June 4-6/74)	1950.68	1	P<.001
No deflector (May 22-24/73)	No deflector (June 11-12/73)	318.97	1	P<.001

[1]Figures in parentheses give the dates of the (pooled) tests considered; these are preceded by the angle of the deflector in relation to the direction of flow, the depth of screens used and indication of whether a sounding lip was utilized.

that efficiency was dependent ($P<0.001$) on the bypass system, but was improved only when a deflector without a sounding lip was used (Tables 2 and 3). The deflector with the sounding lip did not increase guidance efficiency over that obtained for the late control experiments.

Turbine and Bypass Discharges

The relation between turbine discharge, bypass discharge and smolt recapture rate (smolts/hr) for testing conditions with no deflection is illustrated in Figure 6. A significant ($P<0.01$) negative relationship was found between turbine discharge and the rate of smolt recapture; a weak positive correlation ($P<0.1$) was found between bypass discharge and the rate of smolt recapture. The effectiveness of the bypass for directing smolts was diminished with increased turbine discharge, but improved during increased bypass discharge. A similar analysis of data for periods with the deflection screens in use showed no statistically significant correlation between either turbine discharge (1973: $r = -0.080$; 1974: $r = +0.075$) or bypass discharge (1973: $r = +0.027$; 1974: $r = -0.031$) and the rate of smolt recapture. Conditions created by the deflection system apparently masked the effect of variations in discharge on smolt passage rate.

An examination of turbine discharge in relation to bypass discharge during the 1973 and 1974 test periods revealed the strong dependence of bypass flows on turbine discharge ($r = 0.33$; $P<0.001$). This dependence is probably a result of the relatively small storage capacity of the powerhouse headpond and the fixed elevation of the bypass weir.

Tables 4 and 5 illustrate the rate of use of the bypass facility by smolts over the range of flows experienced during the 1973 and 1974 test periods. The majority of smolts in both years, 24.6 to 73.0 percent, used the bypass at the upper level of flow, i.e., 0.32 to 0.53 m^3/sec., and at the fastest rate.

GENERAL DISCUSSION AND CONCLUSIONS

Because of smolt delay in the hydroelectric impoundment, the bypass efficiencies that were deterined are normally under-estimates of the absolute or true efficiency of a test array. Examination of the percentages of the total smolt recoveries taken in the five-day test periods (Tables 1 and 2) revealed that during the early season experiments considerably more test fish were delayed in their descent through the recapture area than in later experiments. This

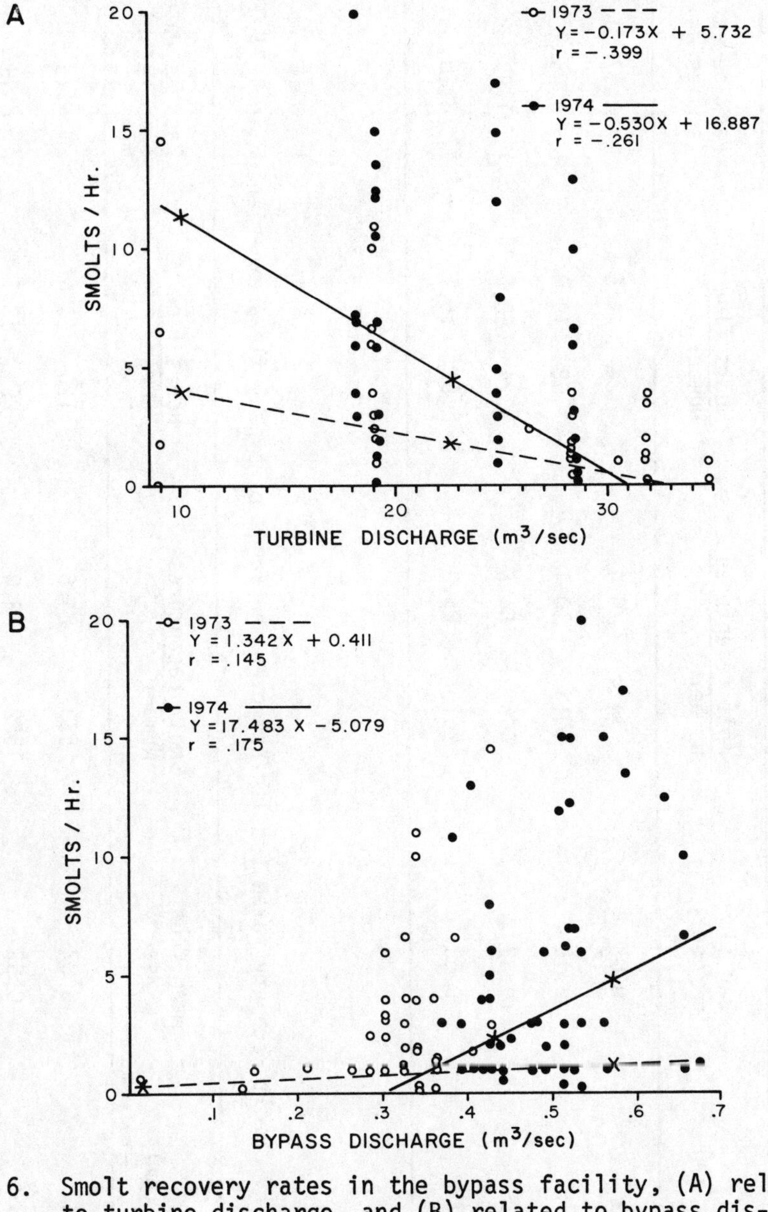

Figure 6. Smolt recovery rates in the bypass facility, (A) related to turbine discharge, and (B) related to bypass discharge for control tests. (Recovery rates of zero were included in the regression analysis, but are not indicated on the graphs.)

Table 4. Smolt recovery rates (smolts/hr) over the range of bypass flows experienced in 1973.

Bypass depth (cm)	Mean flow (m^3/sec.)	Smolt recaptures		Hours flow		Passage rate (smolts/hr)
		Number	%	Number	%	
0-9	0.03	4	0.4	24	4.5	0.17
10-19	0.15	101	10.1	106	19.8	0.95
20-29	0.32	646	64.5	324	60.6	1.99
30-39	0.53	251	25.0	81	15.1	3.10

Table 5. Smolt recovery rates (smolts/hr) over the range of bypass flows experienced in 1974.

Bypass depth (cm)	Mean flow (m^3/sec.)	Smolt recaptures		Hours flow		Passage rate (smolts/hr)
		Number	%	Number	%	
20-29	0.32	523	24.6	126	31.6	4.15
30-39	0.53	1552	73.0	254	63.7	6.11
40-49	0.78	52	2.4	19	4.8	2.74

may have been due to behavioral changes caused by differences between hatchery water temperatures and those at the release site (Fig. 5). Dymond (1963) indicated that the immediate stimulus controlling smolt descent is either a rise in water temperature or a factor associated with rise in temperature.

Some of the test fish may have used the spillway outlet at the diversion dam in 1973, when there was surface discharge on three days (May 18-24). This also would make fewer fish available to the deflector arrays in the test area and result in an underestimate of true guidance efficiency.

Further examination of the numbers of smolts recovered versus time for the early season experiments, revealed that where a more accurate determination of the efficiency of a test array is desired, one should extend the test recovery period for each experiment, until no smolts are taken on three consecutive days after commencement of smolt recapture. Such a precaution yielded only 1-2 test fish recoveries per experiment, a negligible number considering the size of the test lots of smolts released.

The real merit or utility of deflector screens was masked by increased bypass efficiency with progression in the smolt migration season. The reason for this is unclear, but it is undoubtedly related to changes in smolt behavior brought about by hydro-thermal conditions in the river (Fig. 5) and the hydrology at the test site.

An inspection of the bypass efficiency results (Tables 1 and 2) shows that only with certain deflector arrangements can guidance efficiency be increased over that obtained by merely providing a well-positioned bypass with adequate flow.

The high average efficiency of late season control experiments (41.2 to 53.3 percent) may be due partially to the location of the bypass and shape of the river channel leading to the hydroelectric dam. Mills (1965) indicates, with regard to successful smolt passes, that much depends on the correct siting of the pass and the provision of a surface exit. He has shown that where a smolt pass is built at one end of a dam, close to shore, like that at Tusket River, little delay of smolts occurred. At other passes not so sited, considerable delay resulted. It is felt that because of site conditions at the Tusket River hydroelectric dam, smolts are more apt to find the bypass than in a straight run-of-the-river situation.

Because of the significant effect of turbine discharge on the amount of water available for the bypass and the usage of the facility by smolts, it is believed that the efficiency of the bypass can be increased by maintaining the minimum discharge at 0.32 to 0.53 m^3/sec. (Tables 4 and 5). These average bypass discharges resulted in the highest passage rates of smolts and the largest percentages of smolts (24.6 to 73.0 percent) being captured. Flow maintenance in the bypass could be achieved by installing an automatic spill gate which would self-adjust to headpond fluctuations caused by turbine output.

Other factors which should be considered in efficient bypass design are the depth of water at the entrance, and the distance of the bottom lip of the bypass from the water surface. Ducharme (1972) found that nearly 80 percent of Atlantic salmon smolts descending the power canal at East River, Sheet Harbour, Nova Scotia, swam within 130 cm. of the surface. However, only 25 percent were distributed within 45 cm. of the water surface. Observations made in Ireland (McGrath, 1960) and Scotland (Stuart, 1962) also suggest that Atlantic salmon smolts tend to migrate in the surface layers of water and exhibit a preference for a surface outlet. At the Tusket Falls hydroelectric station, smolts were usually observed from near the surface to the limit of light penetration (approximately 250 cm., Secchi disc measurement). These results and those of the present study indicate that the bypass entrance should be located near the water surface. Furthermore, the depth of spill through the bypass should be maintained at a minimum of 30 cm. and possibly be increased to one meter, for optimum passage conditions.

A further possibility for improving the bypass at Tusket River is suggested by Pyefinch (1966) and Stuart (1962). Both authors relate that the provision of an apron slanting downwards and upstream from the pass improves guidance efficiency. The latter contends that this is accomplished by making fish less reluctant to use the outlet because of the presence of visual orientation cues provided by the artificial substratum or sloping ramp.

Average approach velocities (3 to 30 cm/sec., Fig. 4) along the face of the deflector during the Tusket River experiments, are not considered high enough to cause serious impingement of smolts against the screens and detract from their usefulness. Experience has shown that the safe approach velocity for smolts >8cm. long, at screening installations, can be increased to 30 cm/sec. over that which is recommended for smaller fish (Clay, 1961).

Deflection angle proved to be the most important factor for increasing smolt guidance efficiency. By reducing the deflection angle from 90 to 45°, average guidance efficiency was increased by an additional 49.3 percent. Increasing the depth of the deflector screens used increased guidance efficiency by plus 26.5 percent.

The implication of the deflector angle experiments is that if the angle is too high, smolts will sound under the screens and in this instance be more likely to be attracted to the turbine outlets.

It is concluded from the preceding results that smolt guidance efficiency is positively dependent on deeper screening depth and reduced angle of deflection. The most efficient deflector array for smolt guidance of those tested is one in which the deflector screens are operated to a depth of 2.44 m. without a sounding lip and at an angle of 45° to the direction of flow. Under these conditions an average bypass efficiency of 72.1 percent was achieved (Table 2).

LITERATURE CITED

ANDREW, F. J., L. R. KERSEY and P. C. JOHNSON. 1955. An investigation of the problem of guiding downstream-migrant salmon at dams. Internat. Pacific Salmon Fish. Comm., Bull. 8: 1-65.

BATES, D. W. 1970. Diversion and collection of juvenile fish with traveling screens. U. S. Fish and Wildl. Serv., Fish. Leafl. 633: 1-6.

BATES, D. W. and S. G. JEWETT, JR. 1961. Louver efficiency in diverting downstream migrant steelhead. Trans. Amer. Fish. Soc. 90(3): 336-337.

BURNS, J. W. 1966. Fish screens. pp. 156-161. In A. Calhoun (ed.). Inland fisheries management. Calif. Dept. Fish and Game, Resource Agency.

CAMPBELL, H. J. 1959. Field testing rotary fish screens in northeastern Oregon. Prog. Fish-Cult. 21(1): 55-62.

CLAY, C. H. 1961. Design of fishways and other fish facilities. Queen's Printer, Ottawa. 301 pp.

DYMOND, J. R. 1963. Family salmonidae. pp. 457-501. In Y. H. Olsen (ed.). Fishes of the western north Atlantic. Sears Foundation for Marine Research, Memoir No. 1.

DUCHARME, L. J. A. 1972. An application of louvers for guiding Atlantic salmon (*Salmo salar*) smolts from power intakes. J. Fish. Res. Bd. Can. 29(10): 1397-1404.

EICHER, G. J. 1970. Fish passage. pp. 163-171. In N. G. Benson (ed.). A century of fish culture in North America. Amer. Fish. Soc., Spec. Publ. No. 7.

FIELDS, P. E. 1957. Guiding migrant salmon. The Scientific Monthly 25(1): 10-22.

HUBER, E. E. 1974. Fish protection at intake structures and dams: guidance, screens and collection devices. A selected bibliography with abstracts. Oak Ridge National Laboratory, Oak Ridge, Tenn. MS Rept. 75 pp.

KUPKA, K. H. 1966. A downstream migrant diversion screen. Can. Fish-Cult. 37: 27-34.

LEITRITZ, E. 1952. Stopping them: the development of fish screens in California. Calif. Fish and Game 38(1): 53-62.

LETHLEAN, N. G. 1953. An investigation into the design and performance of electric fish-screens and on electric fish-counter. Trans. Roy. Soc. Edin. 62(13) Pt. 2: 479-525.

LONG, C. W. and R. F. KREMA. 1969. Research on a system for bypassing juvenile salmon around low-head dams. Com. Fish. Rev. 21(6): 27-29.

MARQUETTE, W. M. and C. W. LONG. 1971. Laboratory studies of screens for diverting juvenile salmon and trout from turbine intakes. Trans. Amer. Fish. Soc. 100(3): 439-447.

McGRATH, C. J. 1960. Dams as barriers or deterrents to the migration of fish. Proc. Internat. Union Conserv. Nature 4: 81-92.

MILLS, D. H. 1965. Observations on the effects of hydro-electric developments on salmon migration in a river system. I.C.E.S./C.M. No. 32, 5 pp.

MONAN, G. E., J. R. PUGH and J. R. SMITH. 1967. Efficiency of a combined electrode and louver array in guiding juvenile steelhead trout (Salmo gairdneri). Trans. Amer. Fish. Soc. 96(4): 422-423.

PUGH, J. R., G. E. MONAN and J. R. SMITH. 1969. Attempt to guide seaward-migrating Pacific salmon and trout with long leads. Prog. Fish-Cult. 31(3): 174-176.

PYEFINCH, K. A. 1966. Hydro-electric schemes in Scotland, biological problems and effects on salmonid fisheries. pp. 139-147. In Low-McConnel (ed.). Man-made lakes. Academic Press, New York.

REES, W. H. 1957. The vertical and horizontal distribution of seaward migrant salmon in the Forebay of Baker Dam. Wash. Dept. Fish., Fish Res. Pap. 2(1): 5-17.

RUGGLES, C. P. and P. RYAN. 1964. An investigation of louvers as a method of guiding juvenile Pacific salmon. Can. Fish-Cult. 33: 1-68.

SHARMA, R. K. 1973. Fish protection at water diversions and intakes: a bibliography of published and unpublished references. Argonne National Laboratory, Argonne, Ill. MS Rept. 35 pp.

SOKAL, R. R. and F. J. ROHLF. 1969. Biometry. W. H. Freeman and Co., San Francisco. 776 pp.

STUART, T. A. 1962. The leaping behavior of salmon and trout at falls and obstructions. Dept. Agric. Fish. Scot., Freshw. Salmon Fish. Res. 28: 1-46.

WALES, J. H., E. W. MURPHEY and J. HANDLEY. 1950. Perforated plate fish screens. Calif. Fish and Game 36(4): 392-403.

WHALLS, M. J., D. S. SHETTER and J. E. VONDETT. 1957. A simplified rotary fish screen and an automatic water gate. Trans. Amer. Fish. Soc. 86: 371-380.

A LABORATORY STUDY ON THE EFFECTS OF THE EXPOSURE OF SOME ENTRAINABLE HUDSON RIVER BIOTA TO HYDROSTATIC PRESSURE REGIMES CALCULATED FOR THE PROPOSED CORNWALL PUMPED-STORAGE PLANT

Andrew P. Beck, Gerald V. Poje, and William T. Waller
Institute of Environmental Medicine
New York University Medical Center
Sterling Forest, Tuxedo, N.Y. 10016

ABSTRACT

The proposed 2,000 megawatt pumped-storage hydroelectric plant at Cornwall, New York, will withdraw Hudson River biota during the pumping cycle, and deposit them in a 25,000 acre-feet upper reservoir, 1,160 feet above sea level. During periods of peak energy demand the water will return by gravity-flow to the river, turning the hydroelectric turbines and generating power for New York City and the New York Power Pool. One of the potential stresses encountered by entrained organisms will be changes in hydrostatic pressure. The specific pressure regimes encountered during the pumping and generating cycle are quite complex, and will depend on the final design of the facility.

Results of preliminary studies have contributed to the evaluation of how some Hudson River biota may be affected by changes in hydrostatic pressure. These studies led to the development of a programmable pressure chamber, capable of reproducing various patterns of exposure, ranging from 2.0 psia (subatmospheric pressures) to 700 psig. The design of the chamber and its use to determine the importance of variables, such as acclimation pressure, temperature, and the organism's life history stage will be discussed.

The striped bass (*Morone saxatilis*), the amphipod (*Gammarus* spp.), and *Chaoborus* larvae are the three groups of organisms our preliminary studies have investigated. Further investigations of these and other species predominant in the area, will provide information on their survival and behavioral responses to changes in hydrostatic pressure.

INTRODUCTION

Consolidated Edison Company of New York, Inc., has proposed constructing a pumped-storage hydroelectric generating facility at Cornwall, on the Hudson River. The site is located at mile point 55 above the Battery in New York City.

In simple terms the pumped-storage facility is to be used as a peak demand power station. Water from the Hudson River will be pumped to an upper reservoir during periods of low base load power demands, using electricity generated at other facilities. The water will be stored in the upper reservoir until periods of peak power demand, at which time it will be released to return to the river. The energy resulting from the flow of water from the upper reservoir to the river will turn turbines and generate electricity.

The facility as designed, will remove approximately 7.5×10^6 to 9.4×10^7 gpm from the river during the pumping cycle. The estimated time required to fill the upper reservoir is approximately 7 to 8 hours. During the pumping cycle non-screenable organisms inhabiting the water will also be removed. The kinds of organisms entrained will depend on the screening device used. If the classical 3/8 inch square mesh (9.5 mm.) screens are used, the major groups of organisms entrained will be the decomposers (bacteria and fungi), phytoplankton, microzooplankton, macrozooplankton, and fish eggs and larvae. Conversely, during the generating cycle organisms pumped into the upper reservoir, or organisms resident in the upper reservoir may be returned to the river. During the pumping and generating cycles the organisms will be exposed to rapid changes in pressure, velocity shear forces, mechanical buffeting, and turbulence.

New York University has been conducting studies on the effects of one of the entrainment stresses, changes in hydrostatic pressure The results presented in this paper represent the preliminary findings of this research effort.

The effects of the operation of hydroelectric or pumped-storage facilities on fish, as organisms susceptible to being entrained, is of particular interest. The research which has been done on the effects of hydrostatic pressure on fish indicates that the organ or organ system of greatest susceptibility to damage is the gas bladder. In order to appreciate the potential for damage to fish, one must have a basic understanding of the structure of the physiological processes associated with the gas bladder of fish.

Lagler, Bardach, and Miller (1962) state that the gas bladder of fish functions as a hydrostatic organ, an accessory breathing organ, a sound producer, and a resonator in sound perception. In general two major groups of fish can be distinguished based on the structure of the gas bladder, (1) those possessing a gas bladder which is connected to the external medium and (2) those whose gas bladder is isolated from the external medium. Fish possessing an open gas bladder (connected to the external medium) are termed physostomes. Fish possessing a closed gas bladder are termed physoclists.

There are many variations on this simple scheme. Some fish, such as darters, have no gas bladder. In most physostomes the connection of the gas bladder to the external medium occurs through a duct (ductus pneumaticus) connected to the esophagus, while in some clupeids, the duct opens to the gut, and release of gas occurs through an accessory opening near the anus. Many physoclists possessing closed gas bladders as adults have a pneumatic duct in some stages of larval life.

The problems of filling and expelling gas from the bladder are different for physostome and physoclist fish. Most physostome fish regulate the volume of gasses in the bladder by gulping air at the air-water interface. The presence of the ductus pneumaticus makes this possible. Many physostomes can also regulate gas volume by secreting gases into the bladder from the blood stream. Expulsion of gases from the gas bladder is likewise accomplished through the ductus pneumaticus.

In physoclist fish secretion and resorption of gases can only be accomplished through physiological processes. Two parts of the gas bladder wall, together with special blood vessels, make up the gas secreting complex. These are the gas gland and the rete mirable. The two structures permit the concentrating of gases from the blood stream into the bladder at tensions far higher than those that exist in the water around the fish. The resorption of gases from the bladder is accomplished in several ways, but in general, all involve an exchange of gas between the bladder and the blood stream of the fish.

Several significant factors must be remembered in reviewing the effects of hydrostatic pressure on fish.

1. Of particular significance is the behavior of gases under pressure. Boyle's law states that the volume of gas changes inversely with pressure (halve the pressure and double the volume, double the pressure, halve the volume).

2. The rate of adjustment by fish to changes in hydrostatic pressure is dependent on the type of gas bladder possessed by the fish. Physostome fish (open bladder) adjust much more rapidly than physoclist (closed bladder) fish.

3. The ability of fish to tolerate changes in hydrostatic pressure is dependent on the pressure to which the fish is acclimated, the direction of change, the magnitude of change, and the rate of change.

A considerable body of literature exists relating to the effects of hydrostatic pressure changes on aquatic organisms. Most of the pressure studies and reviews are not directly applicable to the pressure regimes expected at Cornwall; however they provide insight into the general realm of conditions and organisms which have been tested. Sleigh and MacDonald (1972), Zimmerman (1970), and Kinne (1972) provide an overall review of the effects of changes in hydrostatic pressure on aquatic organisms.

Fish response to increases in hydrostatic pressure may depend upon the presence of absence of swimbladders. Responses of fish with swimbladders may be different depending on whether the fish are physoclists or physostomes.

Tsvetkov _et al_. (1972) modeled the magnitude and rates of changes in hydrostatic pressure during passage through turbines, and exposed both physostomous and physoclist fishes. To acclimate fish, various pressures (1-6 atmospheres) were applied by air pressure to the surface of the water, and the fish were allowed to adapt to neutral buoyancy. The perch (_Perca fluviatilis_), a physoclist, required 23 to 27 hours to acclimate to neutral buoyancy at one atmosphere. At 3 atmospheres, the required acclimation time was at least 72 hours. The physostomous fishes were unable to acclimate when access to an air-water interface was restricted. The rates of releasing pressure to atmospheric ranged from 0.1 to 6 atm/sec. All surviving fish were maintained 3 to 4 days for observation of possible delayed mortalities.

Pressure-induced mortalities were attributed to swimbladder injury and gas disease. Mortalities occurred within 10 seconds to 15 minutes after the rapid release from the acclimation pressure, when swimbladder damage was involved. The damage incurred by the rapid pressure release was a rupturing of the swimbladder wall. In the physostomous fishes, no swimbladder damage was observed when the rate of pressure release was slowed down. It was observed that rupturing of other internal organs occurred, as a result of the expansion of the swimbladder compressing the other organs.

Roach _Rutilus rutilus_ (L.) fingerlings (a physostome) displayed 100 percent mortality at a pressure release rate of 3 atm/sec., 40 to 72 percent mortality at 0.1 to 0.5 atm/sec. and 10 percent mortality when the rate was below 0.1 atm/sec.

Gas disease was observed when the pressure was released from 2 to 3 atmospheres to atmospheric pressure. Gas disease developed about 20 minutes or more after the pressure was released. Gas disease in fish results when pressure is released and gases become supersaturated in the water and in the organism's blood. Generally, hemorrhaging of blood vessel walls occurs about the eyes, gills, fins and kidneys.

The effects of pressure on macrozooplankton have been noted by several authors (Knight-Jones and Morgan, 1966; Digby, 1972; Enright, 1961, 1962; Ebbecke, 1935, 1944). The literature also contains pressure tolerance data on the amphipod Gammarus and the dipteran Chaoborus, which have species that are endemic in the Hudson River estuary.

Naroska (1968) studied the effects of pressure on the heartbeat rate of the amphipods Gammarus oceanicus and Gammarus duebeni. Initially, Gammarus oceanicus displayed an increase in heartbeat activity at 1,470 psig, followed by a gradual reduction in heartbeat rate. If the high pressure exposure was of short enough duration, rapid decompression was followed by recovery to a normal heartbeat rate. Naroska also determined that lower salinities decreased the pressure tolerance of Gammarus oceanicus.

The larvae of the dipteran Chaoborus exhibit a high sensitivity to rapid changes in hydrostatic pressure due to their special buoyancy organs. Four hydrostatic organs are arranged in pairs on the thorax and the ninth segment. Damant (1924) subjected larvae to various low positive pressures. When exposed to 2.9 psig or 5.8 psig the larvae first exhibited negative buoyancy (they sank to the bottom of their experimental exposure vessel), but after periods of 10 minutes and 60 minutes respectively, they acclimated and attained neutral buoyancy. When exposed to 8.7 psig the larvae failed to acclimate, and remained permanently impaired on the bottom of the exposure vessel. In many cases the semi-rigid wall of the hydrostatic organs ruptured, allowing body fluids to enter the sacs, and the organs appeared normal although in reality they were useless.

MATERIALS AND METHODS

The approach taken in attempting to study the effects of hydrostatic pressure has been to model the pressure profiles calculated to exist at Cornwall, and to expose some representative entrainable organisms to these profiles. None of the various apparatuses used to test the hydrostatic pressure profiles simulates precisely the time-pressure profiles expected to occur.

In the initial phases of the study, we were advised that due to the minus 50 foot submersion of the turbines during the pumping or generating cycles, there would be no pressures less than atmospheric (14.7 psia) in the system. As the study progressed and more detailed analyses were made, it was learned that during both pumping and generating cycles some percentage of the water would pass through a nearly instantaneous reduction in pressure. The magnitude of the change in pressure and the calculated volume of water associated with the reduction in pressure are based on information provided by Consolidated Edison Company of New York and its contractors building the turbines. The methods and results which follow reflect the changes in the interpretation of the pressure regimes expected, and the increased sophistication in experimental apparatuses used to simulate the pressure regimes.

Figures 1 and 2 are schematics of the pressure regimes expected at Cornwall at full power levels. Figure 1 is the schematic of the pumping mode, while Figure 2 represents the generating mode.

Our interpretation of the phenomena concerning pressure changes expected to occur at Cornwall is as follows. A few inches (6" to 8") below the turbine blades there will exist a pressure gradient from subatmospheric to approximately 2 atmospheres of pressure. The minimum pressure condition is approximately 2 psia. Approximately 60 to 70 percent of the entrained water is expected to pass through a zone of 1 to 2 atmospheres of pressure, while 30 to 40 percent of the water is expected to pass through a zone of from 8 psia to atmospheric pressure. Only a small percentage of water is expected to reach 2 psia. These changes in hydrostatic pressure are expected to occur almost instantaneously for any given parcel of water withdrawn, and to occur in both the pumping and generating modes.

The point at which negative pressures are expected to occur in the pumping mode is labeled as B on Figure 1. In the generating mode the point of negative pressure is depicted by the letter K on Figure 2.

During the course of these studies on the effects of hydrostatic pressure on fish and macroinvertebrates, three different types of pressure chambers were used.

The first pressure chamber used in this study was similar to that of Bishai (1961). A schematic of this apparatus is shown in Figure 3. This system was used to expose organisms to pressures less than atmospheric. It could reduce pressures to about 7.5 psia in 2 seconds.

With stopcock "F" closed, and stopcock "C" open, vacuum flask "B" was evacuated and isolated. When the desired pressure was reached stopcock "C" was closed. The test organisms were held in vacuum flask "A" with 1400 ml of water. To reduce the pressures in

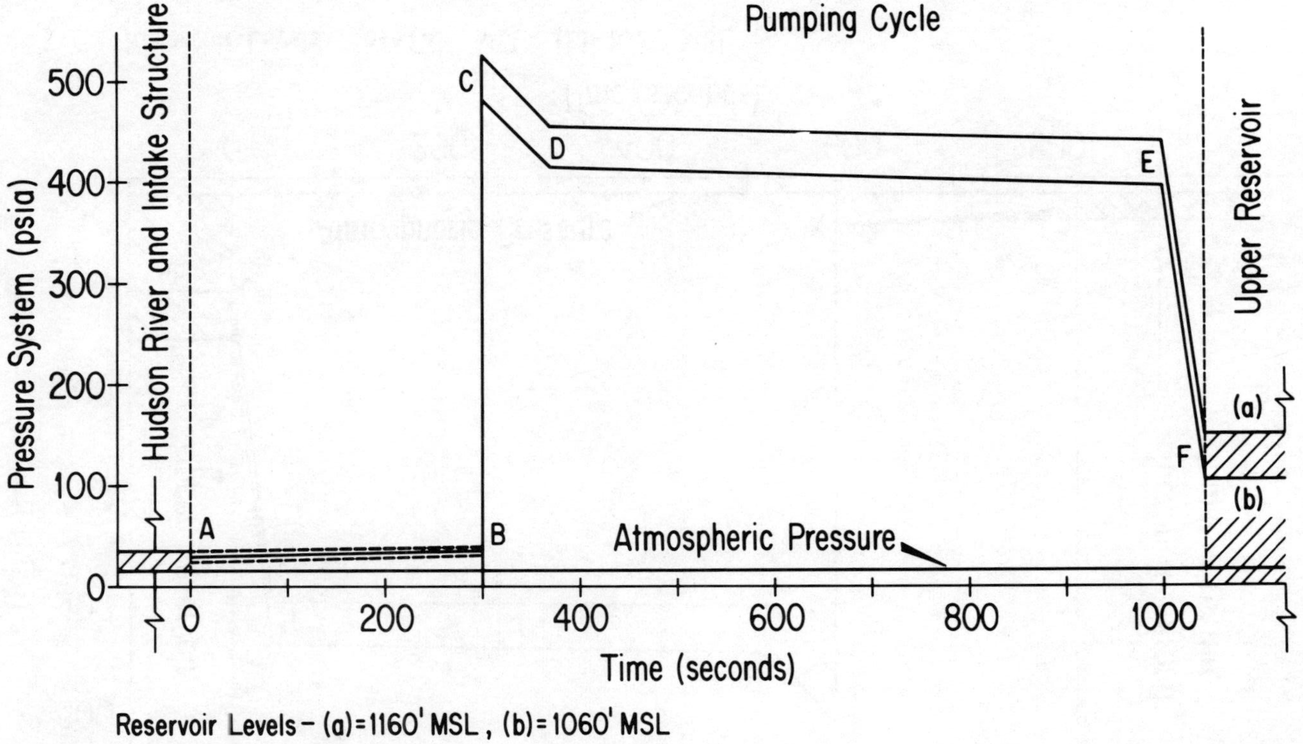

Figure 1. Schematic of pumping cycle pressure regime expected at Cornwall at full power level.

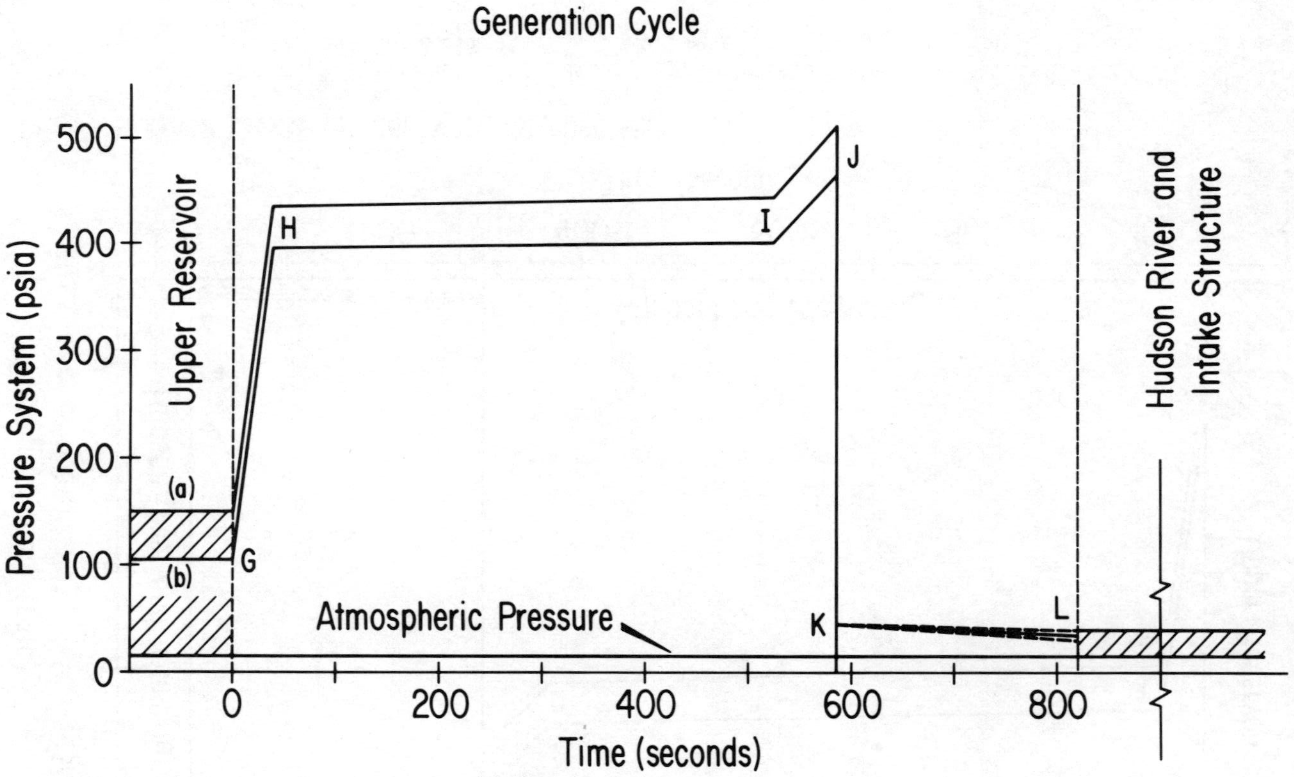

Reservoir Levels — (a) = 1160' MSL, (b) = 1060' MSL

Figure 2. Schematic of generating cycle pressure regime expected at Cornwall at full power level.

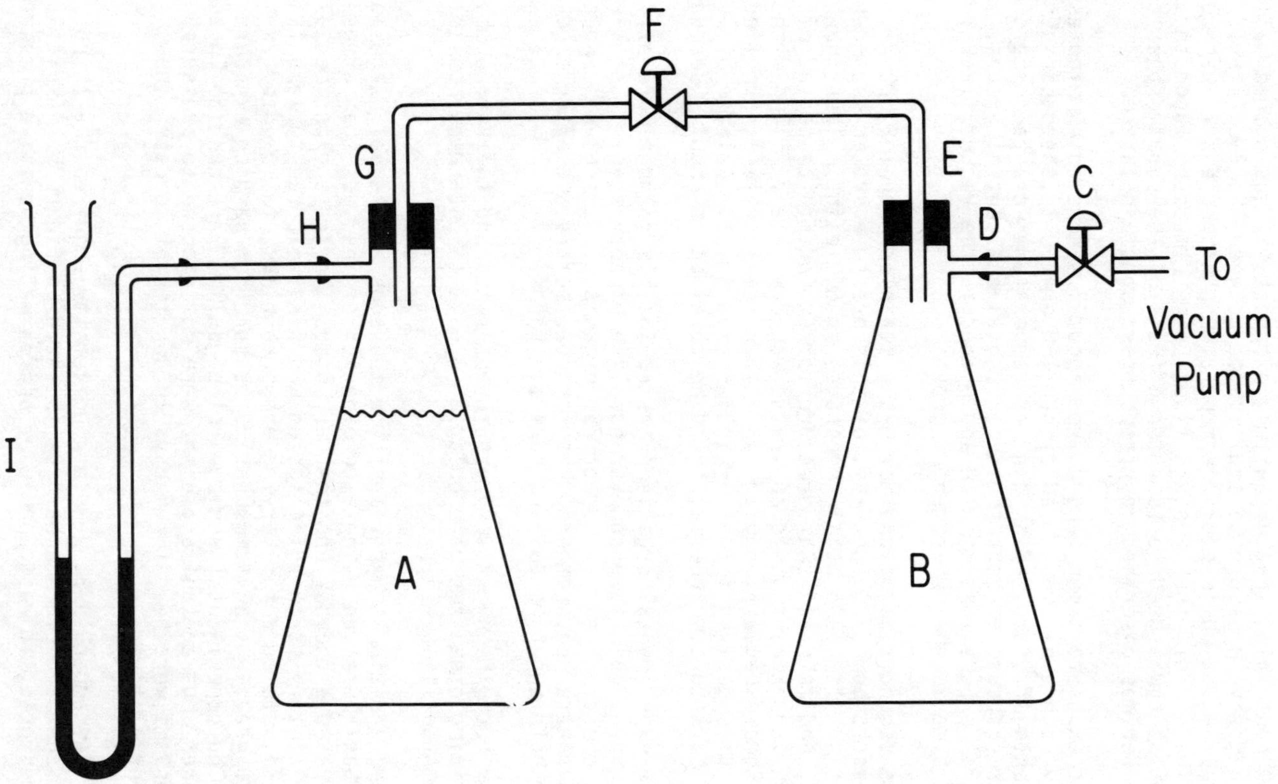

Figure 3. Schematic of the less than atmospheric pressure chamber, similar to that used by Bishai (1961).

the flask containing test organisms, stopcock "F" was opened, allowing the negative pressure to equilibrate between the two flasks. The actual reduction in pressure from atmospheric pressure was recorded directly off the exposure flask using a mercury manometer. This apparatus was used to test the effects of the subatmospheric pressure regimes described on striped bass eggs and larvae.

The exposure times ranged from about 1 second to 60 minutes. Generally each experiment used either 150 or 250 test organisms. Measurements of dissolved oxygen and temperature were taken before and after exposure of both the experimental and control groups. The control groups were handled and placed in the vacuum flask for the same time as the experimental groups.

The second type of chamber was a stainless steel 2,000 ml. Parr pressure bomb. A schematic of this pressure system is shown in Figure 4. It could attain pressures to 800 psig in less than 1 second, and maintain that pressure for an indefinite period of time. During exposure this apparatus was completely filled with water. Compressed air, applied through a small orifice, supplied the pressures desired for a particular experiment. Release of pressure back to atmospheric, could be obtained in about 1 second.

The experimental organisms tested in this chamber were striped bass of entrainable size and various life history stages. Each experiment contained either 200 or 250 organisms. Dissolved oxygen and temperature measurements were recorded before and after exposure for both experimental and control groups.

For both the negative pressure apparatus and the Parr bomb, the surviving organisms were held for 4 days to determine possible delayed mortalities.

Some additional experiments on the eggs and larval stages of striped bass were run using a combination of the less than atmospheric pressure chamber and the Parr bomb. In these experiments the life history stage being tested was first exposed to less than atmospheric pressure for varying exposure times, followed by exposure to greater than atmospheric pressure for varying exposure times. The experimental and control organisms were handled in the same manner, but control organisms were not exposed to pressure. All organisms were held for 4 days in an attempt to determine latent mortalities.

In September 1973, New York University acquired the third pressure chamber. This apparatus is shown in Figure 5. As initially designed, it was capable of approximating positive pressure

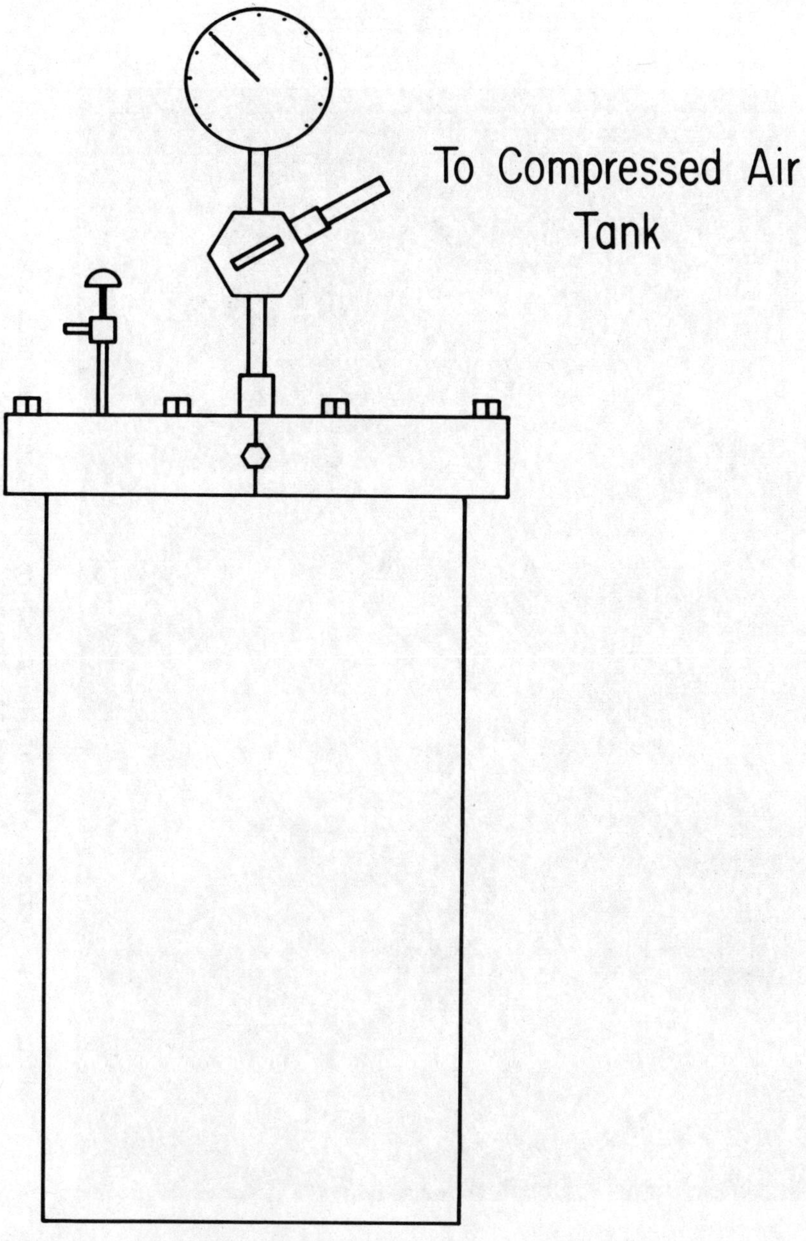

Figure 4. Schematic of the stainless steel, 2,000 ml. Parr pressure bomb.

Figure 5. Overall view of the pressure apparatus designed specifically to simulate the positive and subatmospheric pressure regimes at Cornwall.

regimes similar to those expected to occur at Cornwall. The chamber is equipped with a digital and analog programmer, which sequentially opens and closes ball valves, and in conjunction with two Masoneilan control valves, regulates the amount of pressure and the rate of change of positive pressure in the 227 liter pressure tank. The vessel was constructed prior to receiving the information concerning the less than atmospheric pressures. The pressure vessel in the positive pressure configuration was used to do some preliminary studies on the acclimation of striped bass larvae to positive pressure.

In May 1974, Consolidated Edison modified the pressure vessel so that pressures of less than atmospheric (down to 2 psia) could be obtained. The modification allows for both positive and less than atmospheric pressure exposures.

The modified pressure vessel was used on striped bass eggs, yolk-sac larvae, and larvae as well as some macroinvertebrates, including *Gammarus* species, and *Chaoborus* larvae.

While the pressure regimes to which these organisms were exposed varied somewhat, the majority of experiments was run with organisms acclimated to atmospheric pressure (14.7 psia), followed by exposure to 2 psia followed by exposure to 466 psig, continued through the pumping cycle, and then back to atmospheric. The organisms were held for varying periods of time in an attempt to determine latent mortality.

Vessels for holding test organisms after exposure to pressure to determine delayed mortalities, included ice cube trays, 18 ml. plastic vials, and 600 ml. Mason jars. The bases of the ice cube trays and plastic vials were replaced with 300μ mesh and submerged in quarry water. Others were kept in Mason jars with an air-water interface with circulating quarry water around the jars to maintain them at the ambient temperature. After exposure the organisms were held for periods ranging from 1 day to 34 days, with most experimental and control organisms held for 4 days and then preserved.

Generally, 100 to 250 organisms were used in each experiment. Dependent upon the size and number of test organisms, from 2 to 5 replications were used to reduce crowding effects and the buildup of toxic waste products. In all experiments, the control organisms were handled in the same manner as experimental organisms. This included the means of counting and placing organisms into exposure containers for the same length of time, and then holding the controls for the same period as the experimentals for determination of possible delayed mortalities.

The life history stages of fish used in these experiments were obtained from a variety of sources. All fish, except some of the striped bass life history stages tested, were obtained from the Hudson River. In some cases the stages tested were obtained by seining. For striped bass, artificial propagation of brood stock collected from the Hudson River, provided some life history stages. In addition, a series of experiments on striped bass eggs and yolk-sac larvae were carried out using stages supplied by the Monck's Corner South Carolina Striped Bass Hatchery.[1]

The macroinvertebrates used were obtained from the intakes of the Indian Point steam electric generating complex. Macroinvertebrates were collected using 1/2 meter diameter, #0 mesh nets.

The water in which the pressure tests were carried out depended on the source of the organisms and their immediate past history prior to testing. Some of the striped bass life history stages produced from brood stock taken from the Hudson River were tested in water from a lime stone quarry located at Verplanck, New York. This water was used during the artificial propagation of these stages. Some stages of striped bass as well as the life history stages of other species were tested in either filtered or unfiltered Hudson River water. Those stages of striped bass used from the Monck's Corner Striped Bass Hatchery were tested in well water used by that facility. Macroinvertebrates were tested in both unfiltered Hudson River water and in water from the quarry.

In general the organisms were acclimated to the experimental water for a minimum of 24 hours prior to testing. In most cases acclimation exceeded one week, except for egg stages less than 24 hours old. These stages were tested in the water source in which they had been artificially fertilized.

RESULTS AND DISCUSSION

In all experiments run in these pressure studies the survival of the organisms used was influenced by holding facilities, handling, and natural mortality. In the case of the experimental organisms there existed the potential for additional mortality resulting from exposure to the pressure regimes tested. The immediate

---

[1]The cooperation of Mr. Jack Bayless in providing space for the research and the specimens used at the Monck's Corner Striped Bass Hatchery was greatly appreciated.

observations made after exposure would be least influenced by holding, handling and natural mortality. In all cases tested there existed the potential for the rate of mortality resulting from holding and handling to be so great that it overshadowed latent mortalities due to pressure. These factors should be considered when evaluating latent mortalities, especially for those cases in which control survival dropped below 50 percent. Then analyses were discontinued prior to the completion of the four day holding period.

The results presented in Table 1 are based on the exposure of striped bass eggs, propagated from Hudson River stock, to positive hydrostatic pressures. The time of exposure in each case was 15 minutes, compared to 10-12 minutes projected for the Cornwall Plant. In these experiments there were no significant differences between the survival of control and experimental (exposed) eggs immediately after exposure. Analysis was by contingency tables (α = 0.05). In all cases analyses were discontinued when control survival dropped below 50 percent. Eggs were considered dead when there was evidence under microscopic examination of any of the following abnormalities: discoloration, ruptured chorion, and/or ruptured oil globule or yolk. The mean initial and final temperatures for all experiments combined in Table 1 were the same, 16.0°C, with a range of 14.3°C to 18.2°C.

In addition to the experiments on striped bass eggs, Table 1 shows some results obtained from the exposure of striped bass larvae acclimated to atmospheric pressure followed by exposure for 15 minutes to the pressures indicated. In this series of experiments there was a significant difference (contingency table analysis, α = 0.05) between control and experimental survival for 5 day old striped bass larvae exposed to 715 psig for 15 minutes (note that maximum projected pressure for the Cornwall plant is 510 psig). The difference in immediate survival between control and experimental groups for this larval stage was 14 percent. None of the other larval stages tested in this series of experiments showed any significant immediate mortality. For those cases in which observations for latent survival could be made, no significant mortality was detected.

The results presented in Table 2 are based on the exposure of striped bass eggs and larvae to pressures less than atmospheric. Organisms used in these experiments were tested in the pressure chamber shown in Figure 3. Prior to exposure the organisms had been held at atmospheric pressure. The eggs exposed in these experiments were from Hudson River brood fish while the larvae were from Monck's Corner. The mean initial and final temperatures for both control and experimental groups of larvae were the same,

Table 1. Results of exposure of striped bass eggs and larvae to positive pressure increases.

Stage	(psig) Pressure Initial	Pressure Final	Immediate C % al	E % al	Sig.	1 day C % al	E % al	Sig.
Eggs								
3 hrs.	725	720	96.8	95.2	N.S.	84.8	86.4	N.S.
4 hrs.	525	525	96.8	96.8	N.S.	84.8	76.8	N.S.
25 hrs.	720	710	92.8	93.6	N.S.			
26 hrs.	530	530	92.8	96.0	N.S.			

Stage	(psig) Pressure Initial	Pressure Final	Immediate C % al	E % al	Sig.	1 day C % al	E % al	Sig.	2 day C % al	E % al	Sig.	4 day C % al	E % al	Sig.
Larvae														
9 days 12 hrs.	630	30	99.2	100.0	N.S.									
15 days 16 hrs.	740	500	100.0	99.2	N.S.	95.2	99.2	N.S.				53.6	50.4	N.S.
16 days 10 hrs.	790	510	100.0	100.0	N.S.							80.8	44.8	N.S.
21 days 13 hrs.	1240	1020	100.0	98.4	N.S.				89.6	90.4	N.S.			
50 days	715	715	100.0	86.0	*									

Table 2. Results of exposure of striped bass eggs and larvae to pressures less than atmospheric.

Stage	Exposure (psia)	Exposure time	Immediate C % al	Immediate E % al	Immediate Sig.	1 day C % al	1 day E % al	1 day Sig.
Eggs								
4 hrs.	5.7	15 sec.	96.2	97.6	N.S.	92.0	82.4	*
25 hrs.	5.9	10 sec.	92.8	96.0	N.S.			

Stage	Exposure (psia)	Exposure time	Immediate C % al	Immediate E % al	Immediate Sig.	1 day C % al	1 day E % al	1 day Sig.	3 day C % al	3 day E % al	3 day Sig.
Larvae											
4 days 8 hrs.	10.1	10 sec.	100.0	100.0	N.S.						
4 days 10 hrs.	5.6	5 sec.	100.0	80.0	*	67.2	35.2	*			
5 days	5.6	5 sec.	100.0	100.0	N.S.	44.8	48.0	N.S.			
5 days 7 hrs.	6.7	5 sec.	100.0	100.0	N.S.						
7 days 12 hrs.	6.1	3 sec.	99.2	99.2	N.S.	73.6	53.6	*	54.4	32.8	*
8 days 12 hrs.	6.2	5 sec.	99.2	100.0	N.S.	76.0	77.6	N.S.			
17 days 16 hrs.	8.9	10 sec.	98.4	100.0	N.S.	88.0	94.0	N.S.			

17.8°C, with a range of 17.3°C to 20.5°C. For all experiments with eggs, the temperature remained constant at 14.9°C.

In these experiments 4 and 25 hour old striped bass eggs showed no significant mortality as determined by contingency table analysis (α = 0.05), when observed immediately after exposure. However, after 24 hours holding, a significant difference in survival of 4 hour eggs was detected. In this case the difference in control and experimental survival was 9.6 percent.

The larval experiments reported in Table 2 show that based on immediate observations only one stage (4 day + 10 hour old) of larvae showed a significant difference between control and experimental survival. This statistically significant difference was also recorded 1 day after exposure. By the end of 1 day after exposure a significant difference in the survival of 7 day + 12 hour old larvae was also detected. Subsequently, on the third day after holding, the 7 day + 12 hour old larvae continued to show a significant difference in control and experimental survival.

The experiments summarized in Table 3 are based on the exposure of striped bass eggs artificially propagated at the Monck's Corner Striped Bass Hatchery. These experiments were carried out using the pressure chamber shown in Figure 4. In these experiments one group of 50 eggs, C(1), served as the control for 3 experimental groups of 50 eggs each. One experimental group was exposed to 250 psig, E(2), another to 500 psig, E(3) (approximately the maximum expected in the Cornwall plant) while the last group was exposed to 700 psig, E(4). The exposure time in all cases was 15 minutes. Observations on the survival of these eggs were made at 8 hour intervals for a 96 hour period. The combined mean temperatures for both initial and final temperatures was 18.7°C. The mean initial dissolved oxygen value was 7.9 mg/l, ranging from 7.1 mg/l to 8.6 mg/l; while the mean final dissolved oxygen value was 7.8 mg/l, ranging from 7.0 mg/l to 8.4 mg/l.

The analysis of egg survival immediately after exposure indicated that in both the 1 and 12 hour egg stages the control survival was significantly greater than that for eggs exposed to 250 psig, but was not significantly different from that for eggs exposed to 500 or 700 psig. By the 16 hour observation period the control survival of 1 hour and 12 hour eggs was less than 50 percent and analyses were discontinued.

By the end of the 8 hour observation period the survival of 1 hour control eggs was not different from the previously different 250 psig experimental group, nor was it different from the survival of the 700 psig experimental group. However, the control

Table 3. Results of exposure of various striped bass egg stages to 250, 500 or 700 psig.

	0 hrs.					8 hrs.					16 hrs.				
Stage	C(1) % al	E(2) % al	E(3) % al	E(4) % al	Sig.	C(1) % al	E(2) % al	E(3) % al	E(4) % al	Sig.	C(1) % al	E(2) % al	E(3) % al	E(4) % al	Sig.
Eggs															
1 hr.	98	86	100	98	1>2	70	66	86	60	3>1					
12 hrs.	94	80	86	98	1>2	64	48	70	86	4>1,1>2					
24 hrs.	100	100	100	100	N.S.	96	98	100	100	N.S.	96	98	100	100	N.S.
41 hrs.	100	100	100	100	N.S.	100	100	100	100	N.S.	100	100	100	100	N.S.
	24 Hrs.					32 hrs.					40 hrs.				
Eggs															
1 hr.															
12 hrs.															
24 hrs.	96	98	100	100	N.S.	96	98	100	100	N.S.	96	98	100	100	N.S.
41 hrs.	98	97	98	98	N.S.	88	82	94	84	N.S.	88	78	88	84	N.S.
	48 hrs.					56 hrs.					64 hrs.				
Eggs															
1 hr.															
12 hrs.															
24 hrs.	84	76	90	90	N.S.	84	76	90	90	N.S.	82	76	84	86	N.S.
41 hrs.	88	78	86	84	N.S.	88	76	86	82	N.S.	88	74	86	80	N.S.
	72 hrs.					80 hrs.					88 hrs.				
Eggs															
1 hr.															
12 hrs.															
24 hrs.	80	76	80	82	N.S.	76	76	80	80	N.S.	76	76	78	76	N.S.
41 hrs.	88	74	84	80	N.S.	88	74	84	80	N.S.	88	74	84	80	N.S.
	96 hrs.														
Eggs															
1 hr.															
12 hrs.															
24 hrs.	74	76	78	76	N.S.										
41 hrs.	88	74	84	80	N.S.										

survival was found to be significantly less than that of the experimental group exposed to 500 psig.

By the end of the 8 hour observation period the survival of the control group for 12 hour old eggs continued to be significantly greater than that of the experimental group exposed to 250 psig, but was significantly less than that of the survival recorded for the experimental group exposed to 700 psig.

No significant differences in survival between the control group of either the 24 hour or 41 hour old eggs and the experimental groups tested were found in any of the observation periods.

Striped bass larvae from 1 hour after hatch to 4 days old were exposed to the experimental regime just outlined for eggs. The larvae were hatched from eggs artificially propagated at the Monck's Corner Striped Bass Hatchery (Table 4). These experiments further demonstrate the difficulties in interpreting data on latent effects based on statistical analysis sensitive to small changes in survival.

Through 16 hours of observation after exposure no significant differences in control and experimental survival were detected. At 24 hours the control survival of 1 hour old larvae was determined to be greater than that observed for larvae exposed to 250 psig, E(2), but not different from survival of larvae exposed to 500 psig, E(3), or 700 psig, E(4). This trend continued through the 40 hour observation period, but by 40 hours the survival of the control was also significantly greater than that of larvae exposed to 500 psig. This trend continued until the 64 hour observation period. At 64 hours no significant differences in survival between the control group and any of the experimental groups were detected. This trend continued throughout the remainder of the observations made through 96 hours.

This type of apparent anomaly was observed in other experiments. While in a strict statistical sense there were no differences in survival detected at 64 hours and for subsequent observations for 1 hour old larvae, the statistical difference between accepting or rejecting a difference in survival was so small that some allowance for interpretation should probably be made in these cases.

The results presented in Table 5 are based on the exposure of striped bass eggs taken from atmospheric pressure to less than atmospheric pressure, 6.4 psia, E(2), and 9.2 psia, E(3), and returned to atmospheric conditions in less than 1 second (instantaneously). Table 5 also shows a second experiment with the same

Table 4. Results of exposure of various striped bass larval stages to 250, 500, or 700 psig.

Stage	0 hrs.					8 hrs.					16 hrs.				
	C(1) % al	E(2) % al	E(3) % al	E(4) % al	Sig.	C(1) % al	E(2) % al	E(3) % al	E(4) % al	Sig.	C(1) % al	E(2) % al	E(3) % al	E(4) % al	Sig.
Larvae															
1 hr.	100	100	100	100	N.S.	100	100	100	100	N.S.	100	100	100	100	N.S.
1 day	100	100	100	100	N.S.	100	100	100	100	N.S.	100	100	100	100	N.S.
2 days	100	100	100	98	N.S.	98	100	100	98	N.S.	98	100	100	98	N.S.
3 days	100	100	100	100	N.S.	100	100	100	100	N.S.	100	96	98	98	N.S.
4 days	100	100	100	100	N.S.	100	100	100	100	N.S.	100	100	100	100	N.S.
	24 hrs.					32 hrs.					40 hrs.				
Larvae															
1 hr.	100	82	92	96	1>2	100	82	90	94	1>2	98	80	84	92	1>2,3
1 day	100	100	100	100	N.S.	98	100	100	98	N.S.	98	100	100	98	N.S.
2 days	98	100	100	98	N.S.	94	98	100	98	N.S.	88	62	58	68	1>2,3,4
3 days	90	90	98	90	N.S.	90	88	96	92	N.S.	88	86	94	86	N.S.
4 days															
	48 hrs.					56 hrs.					64 hrs.				
Larvae															
1 hr.	96	78	80	88	1>2,3	94	74	80	86	1>2,3	92	72	78	82	N.S.
1 day	74	42	82	80	1>2	74	42	82	80	1>2	72	38	76	72	1>2
2 days	78	56	56	62	N.S.	78	56	56	62	N.S.	64	46	42	48	N.S.
3 days	88	86	94	86	N.S.	78	70	74	74	N.S.	72	60	62	62	N.S.
4 days															
	72 hrs.					80 hrs.					88 hrs.				
Larvae															
1 hr.	88	70	74	78	N.S.	88	70	74	78	N.S.	86	68	74	74	N.S.
1 day	52	26	50	36	1>2										
2 days															
3 days	64	38	50	56	N.S.	62	38	46	50	N.S.	60	34	46	46	N.S.
4 days															
	96 hrs.														
Larvae															
1 hr.	86	68	74	74	N.S.										
1 day															
2 days															
3 days															
4 days															

Table 5. Results of exposure of striped bass egg stages to less than atmospheric pressures, 6.4 psia or 9.2 psia.

Stage	Immediate				8 hrs.				16 hrs.			
	C(1) % al	E(2) % al	E(3) % al	Sig.	C(1) % al	E(2) % al	E(3) % al	Sig.	C(1) % al	E(2) % al	E(3) % al	Sig.
Instantaneous exposure												
Egg 1 hr.	100	100	100	N.S.	98	98	96	N.S.	98	98	96	N.S.
Egg 12 hrs.	100	100	100	N.S.	100	100	100	N.S.	100	100	100	N.S.
Egg 24 hrs.	100	100	100	N.S.	98	98	98	N.S.	98	96	98	N.S.
Egg 48 hrs.	100	100	100	N.S.	100	100	100	N.S.	98	98	98	N.S.
Instantaneous exposure	24 hrs.				32 hrs.				40 hrs.			
Egg 1 hr.	76	78	80	N.S.	50	38	46	N.S.				
Egg 12 hrs.	100	96	98	N.S.	96	78	74	N.S.	78	66	64	N.S.
Egg 24 hrs.	98	94	98	N.S.	98	92	90	N.S.	94	84	86	N.S.
Egg 48 hrs.	98	98	98	N.S.	84	94	94	N.S.	74	92	94	1>2,3
Instantaneous exposure	48 hrs.				56 hrs.				64 hrs.			
Egg 1 hr.												
Egg 12 hrs.	74	54	52	N.S.	68	50	44	N.S.	68	50	44	N.S.
Egg 24 hrs.	92	74	78	1>2	90	74	76	N.S.	84	66	70	N.S.
Egg 48 hrs.	74	86	86	N.S.	74	80	88	N.S.	70	80	84	N.S.
Instantaneous exposure	72 hrs.				80 hrs.				88 hrs.			
Egg 1 hr.												
Egg 12 hrs.	66	50	42	N.S.	66	50	42	N.S.	64	50	42	N.S.
Egg 24 hrs.	82	60	66	1>2	80	58	62	1>2	78	56	60	1>2
Egg 48 hrs.	70	76	76	N.S.	68	74	76	N.S.	66	70	74	N.S.
Instantaneous exposure	96 hrs.											
Egg 1 hr.												
Egg 12 hrs.	64	50	42	N.S.								
Egg 24 hrs.	78	54	58	1>2								
Egg 48 hrs.												

Table 5, continued

Stage	Immediate C(1) % al	E(2) % al	E(3) % al	Sig.	8 hrs. C(1) % al	E(2) % al	E(3) % al	Sig.	16 hrs. C(1) % al	E(2) % al	E(3) % al	Sig.
10 sec. expos.												
Egg 1 hr.	100	100	100	N.S.	98	96	94	N.S.	98	96	94	N.S.
Egg 12 hrs	100	100	100	N.S.	100	100	100	N.S.	100	100	100	N.S.
Egg 24 hrs.	100	100	98	N.S.	100	100	98	N.S.	100	100	98	N.S.
Egg 48 hrs.	100	100	100	N.S.	98	98	100	N.S.	98	98	100	N.S.
	24 hrs.				32 hrs.				40 hrs.			
10 sec. expos.												
Egg 1 hr.	74	74	72	N.S.	60	54	52	N.S.	56	46	48	N.S.
Egg 12 hrs.	96	90	86	N.S.	86	82	78	N.S.	78	70	70	N.S.
Egg 24 hrs.	94	98	98	N.S.	94	94	96	N.S.	92	88	90	N.S.
Egg 48 hrs.	80	88	94	N.S.	80	88	90	N.S.	80	86	88	N.S.
	48 hrs.				56 hrs.				64 hrs.			
10 sec. expos.												
Egg 1 hr.	52	42	44	N.S.	50	38	40	N.S.	50	36	38	N.S.
Egg 12 hrs.	72	66	56	N.S.	60	56	48	N.S.	60	54	46	N.S.
Egg 24 hrs.	86	82	84	N.S.	82	78	80	N.S.	76	74	72	N.S.
Egg 48 hrs.	78	82	82	N.S.	78	80	80	N.S.	76	76	76	N.S.
	72 hrs.				80 hrs.				88 hrs.			
10 sec. expos.												
Egg 1 hr	50	36	38	N.S.								
Egg 12 hrs.	58	52	44	N.S.	58	48	42	N.S.	56	46	42	N.S.
Egg 24 hrs.	74	68	66	N.S.	70	62	66	N.S.	70	62	60	N.S.
Egg 48 hrs.	74	72	72	N.S.	74	64	68	N.S.	72	64	68	N.S.
	96 hrs.											
10 sec. expos.												
Egg 1 hr.												
Egg 12 hrs.	56	46	40	N.S.								
Egg 24 hrs.	68	62	52	N.S.								
Egg 48 hrs.	72	60	62	N.S.								

same exposure pattern except that the eggs were maintained at the less than atmospheric pressure indicated for 10 seconds prior to being returned to atmospheric pressure. The means of initial and final temperatures for experiments in Tables 5 and 6 were the same, 18.7°C. The mean initial and final dissolved oxygen values were 7.1 mg/l, ranging from 7.0 mg/l to 7.4 mg/l (in both initial and final measurements).

These data indicate that for the egg stages tested and the pressure regimes used, except possibly the 24 hour egg exposed to 6.4 psia, survival should be very good. Similar high survival values were found for larvae exposed to the same experimental regime. These data are summarized in Table 6.

The results summarized in Tables 7 and 8 are based on the exposure of striped bass eggs and larvae taken from atmospheric conditions to a less than atmospheric pressure of 6.4 psia, followed by exposure to atmospheric pressure and subsequent exposure to either 250 psig, E(2), 500 psig, E(3), or 700 psig, E(4), for 15 minutes prior to being returned to atmospheric conditions. A combination of the experimental chambers shown in Figures 3 and 4 was used in these experiments. Approximately 2 minutes were required to transfer the organisms to the positive pressure test vessel after exposure to negative pressure. The mean values for initial and final temperatures in Tables 7 and 8 were the same, 18.7°C. The mean values for both initial and final dissolved oxygen measurements were 7.1 mg/l, ranging from 7.0 mg/l to 8.0 mg/l. For all stages tested no significant difference in control and experimental survival was detected.

The results given in Table 9 are based on the exposure of various life history stages of striped bass in the pressure chamber shown in Figure 5. In these experiments the life history stages tested were removed from the holding facilities at approximately atmospheric conditions and transferred to the pressure chamber. The pressure was then reduced to 2 psia in approximately 2 seconds. After reaching 2 psia the pressure was returned to atmospheric in 2 to 3 seconds. After reaching atmospheric pressure there was a 10 second delay prior to exposure to 466 psig. Once 466 psig was attained, the time pressure exposure followed that is shown schematically in Figure 1. Once the pressure cycle reached point F (the termiantion of the pumping cycle) the organisms were returned to atmospheric pressure. The mean initial and final temperature values for the controls and experimentals in Table 9 were the same; i.e., 17.7°C, ranging from 17.1°C to 20.4°C. The mean initial and final dissolved oxygen values for the controls were 11.1 mg/l, ranging from 10.8 mg/l to 11.4 mg/l. The mean initial dissolved oxygen value for the experimental group was

Table 6. Results of exposure of striped bass larval stages to less than atmospheric pressures, 6.4 psia or 3.2 psia.

Stage	Immediate C(1) % al	Immediate E(2) % al	Immediate E(3) % al	Sig.	8 hrs. C(1) % al	8 hrs. E(2) % al	8 hrs. E(3) % al	Sig.	16 hrs. C(1) % al	16 hrs. E(2) % al	16 hrs. E(3) % al	Sig.
Instantaneous exposure												
Larvae 1 hr.	100	100	100	N.S.	94	94	96	N.S.	86	82	94	N.S.
Larvae 1 day	100	100	100	N.S.	82	96	96	N.S.	64	96	96	1>2,3
Larvae 2 days	100	100	100	N.S.	96	100	100	N.S.	96	100	100	N.S.
Larvae 3 days	100	100	100	N.S.	100	100	100	N.S.	100	100	100	N.S.
Larvae 4 days	100	100	100	N.S.	100	100	100	N.S.	100	100	100	N.S.
Instantaneous exposure	24 hrs.				32 hrs.				40 hrs.			
Larvae 1 hr.	82	78	86	N.S.	82	74	82	N.S.	82	70	78	N.S.
Larvae 1 day	64	86	84	N.S.	64	74	72	N.S.	64	66	68	N.S.
Larvae 2 days	96	98	94	N.S.	92	94	96	N.S.	90	92	88	N.S.
Larvae 3 days	100	100	100	N.S.	98	96	98	N.S.	98	94	96	N.S.
Larvae 4 days	100	100	100	N.S.	100	96	98	N.S.	100	94	96	N.S.
Instantaneous exposure	48 hrs.				56 hrs.				64 hrs.			
Larvae 1 hr.	82	70	78	N.S.	80	70	78	N.S.	78	66	76	N.S.
Larvae 1 day	62	56	56	N.S.	60	50	54	N.S.	60	48	52	N.S.
Larvae 2 days	90	88	84	N.S.	88	88	84	N.S.	88	86	80	N.S.
Larvae 3 days	96	92	94	N.S.	94	82	86	N.S.	86	68	80	N.S.
Larvae 4 days	98	94	90	N.S.	90	86	82	N.S.	88	78	82	N.S.
Instantaneous exposure	72 hrs.				80 hrs.				88 hrs.			
Larvae 1 hr.	76	64	72	N.S.	74	62	70	N.S.	74	62	70	N.S.
Larvae 1 day	60	48	52	N.S.	58	48	52	N.S.	58	46	50	N.S.
Larvae 2 days	88	84	80	N.S.	88	84	80	N.S.	88	82	78	N.S.
Larvae 3 days	86	64	78	N.S.	82	64	70	N.S.	80	64	70	N.S.
Larvae 4 days	88	76	80	N.S.	84	76	78	N.S.	82	72	74	N.S.
Instantaneous exposure	96 hrs.											
Larvae 1 hr.	72	60	68	N.S.								
Larvae 1 day	58	46	50	N.S.								
Larvae 2 days	88	82	78	N.S.								
Larvae 3 days	80	62	70	N.S.								
Larvae 4 days	82	72	74	N.S.								

Table 6, continued

Stage	Immediate				8 hrs.				16 hrs.			
	C(1) % al	E(2) % al	E(3) % al	Sig.	C(1) % al	E(2) % al	E(3) % al	Sig.	C(1) % al	E(2) % al	E(3) % al	Sig.
10 sec. expos.												
Larvae 1 hr.	100	100	100	N.S.	94	98	98	N.S.	82	90	90	N.S.
Larvae 1 day	100	100	100	N.S.	94	98	98	N.S.	88	98	98	N.S.
Larvae 2 days	100	100	100	N.S.	100	100	100	N.S.	100	100	100	N.S.
Larvae 3 days	100	100	100	N.S.	100	100	100	N.S.	100	98	100	N.S.
Larvae 4 days	100	100	100	N.S.	100	100	100	N.S.	100	100	100	N.S.
10 sec. expos.	24 hrs.				32 hrs.				40 hrs.			
Larvae 1 hr.	80	86	86	N.S.	78	82	84	N.S.	78	82	80	N.S.
Larvae 1 day	86	90	84	N.S.	84	82	74	N.S.	84	74	66	N.S.
Larvae 2 days	100	94	94	N.S.	98	92	94	N.S.	98	92	94	N.S.
Larvae 3 days	100	98	98	N.S.	98	96	96	N.S.	94	98	92	N.S.
Larvae 4 days	98	98	100	N.S.	98	96	100	N.S.	98	92	96	N.S.
10 sec. expos.	48 hrs.				56 hrs.				64 hrs.			
Larvae 1 hr.	78	82	78	N.S.	76	80	74	N.S.	76	80	74	N.S.
Larvae 1 day	84	64	60	1>2,3	78	60	58	N.S.	78	58	58	N.S.
Larvae 2 days	98	92	90	N.S.	98	88	80	1>3	96	84	78	1>3
Larvae 3 days	90	80	80	N.S.	90	64	68	1>2,3	78	60	58	N.S.
Larvae 4 days	98	92	94	N.S.	96	88	86	N.S.	96	82	86	N.S.
10 sec. expos.	72 hrs.				80 hrs.				88 hrs.			
Larvae 1 hr.	74	76	72	N.S.	74	76	70	N.S.	74	74	70	N.S.
Larvae 1 day	78	56	56	1>2,3	78	56	56	1>2,3	76	56	56	N.S.
Larvae 2 days	94	84	78	N.S.	94	84	78	N.S.	94	84	78	N.S.
Larvae 3 days	78	58	58	N.S.	78	58	58	N.S.	74	56	56	N.S.
Larvae 4 days	94	80	84	N.S.	92	80	78	N.S.	90	78	76	N.S.
10 sec. expos.	96 hrs.											
Larvae 1 hr.	74	74	68	N.S.								
Larvae 1 day	76	56	56	N.S.								
Larvae 2 days	94	84	78	N.S.								
Larvae 3 days	72	54	54	N.S.								
Larvae 4 days	90	76	76	N.S.								

Table 7. Results of exposure of striped bass egg stages to less than atmospheric pressure (6.4 psia), followed by exposure to atmospheric pressure, and subsequent exposure to either 250, 500, or 700 psig.

	0 hrs.					8 hrs.					16 hrs.				
Stage	C(1) % al	E(2) % al	E(3) % al	E(4) % al	Sig.	C(1) % al	E(2) % al	E(3) % al	E(4) % al	Sig.	C(1) % al	E(2) % al	E(3) % al	E(4) % al	Sig.
Eggs															
1 hr.	98	94	96	94	N.S.	78	74	74	78	N.S.	78	72	74	78	N.S.
12 hrs.	98	98	100	98	N.S.	98	98	100	98	N.S.	92	88	88	96	N.S.
24 hrs.	100	100	100	100	N.S.	100	100	100	100	N.S.	100	100	100	100	N.S.
41 hrs.	100	100	100	100	N.S.	98	100	98	100	N.S.	98	100	98	100	N.S.
	24 hrs.					32 hrs.					40 hrs.				
Eggs															
1 hr.	72	70	66	62	N.S.	54	54	54	60	N.S.	50	46	36	40	N.S.
12 hrs.	84	74	78	82	N.S.	80	70	66	74	N.S.	66	62	60	58	N.S.
24 hrs.	98	94	90	92	N.S.	90	82	78	86	N.S.	74	70	66	76	N.S.
41 hrs.	98	94	96	98	N.S.	92	86	94	98	N.S.	90	82	92	84	N.S.
	48 hrs.					56 hrs.					64 hrs.				
Eggs															
1 hr.															
12 hrs.	62	54	54	56	N.S.	60	50	52	52	N.S.	58	50	50	46	N.S.
24 hrs.	74	62	56	66	N.S.	66	52	54	58	N.S.	66	52	54	52	N.S.
41 hrs.	90	78	88	82	N.S.	84	78	82	82	N.S.	84	78	82	80	N.S.
	72 hrs.					80 hrs.					88 hrs.				
Eggs															
1 hr.															
12 hrs.	54	46	46	42	N.S.	52	42	46	42	N.S.	52	40	44	40	N.S.
24 hrs.	66	50	52	52	N.S.	62	48	48	50	N.S.	58	48	46	50	N.S.
41 hrs.	76	72	80	76	N.S.	76	70	74	72	N.S.	76	70	70	70	N.S.
	96 hrs.														
Eggs															
1 hr.															
12 hrs.	52	40	42	40	N.S.										
24 hrs.	58	46	46	50	N.S.										
41 hrs.	76	68	70	70	N.S.										

Table 8. Results of exposure of striped bass larval stages to less than atmospheric pressure (6.4) psia), followed by exposure to atmospheric pressure, and subsequent exposure to either 250, 500, or 700 psig.

Stage	0 hrs.					8 hrs.					16 hrs.				
	C(1) % al	E(2) % al	E(3) % al	E(4) % al	Sig.	C(1) % al	E(2) % al	E(3) % al	E(4) % al	Sig.	C(1) % al	E(2) % al	E(3) % al	E(4) % al	Sig.
Larvae															
2 hrs.	100	100	100	100	N.S.	100	100	100	100	N.S.	98	94	96	94	N.S.
1 day	100	100	100	100	N.S.	100	98	100	98	N.S.	98	88	94	92	N.S.
2 days	100	100	100	100	N.S.	100	100	100	100	N.S.	100	100	100	100	N.S.
3 days	100	100	98	100	N.S.	100	100	96	100	N.S.	100	100	96	98	N.S.
4 days	100	100	100	100	N.S.	100	100	100	100	N.S.	100	100	98	100	N.S.
Larvae	24 hrs.					32 hrs.					40 hrs.				
2 hrs.	98	94	92	90	N.S.	96	94	92	82	N.S.	94	88	90	80	N.S.
1 day	94	78	82	88	N.S.	92	78	84	82	N.S.	90	76	84	82	N.S.
2 days	100	100	100	100	N.S.	98	92	98	90	N.S.	96	92	94	86	N.S.
3 days	98	98	94	94	N.S.	96	92	90	86	N.S.	90	92	88	82	N.S.
4 days	100	100	98	100	N.S.	98	100	98	98	N.S.	96	98	96	94	N.S.
Larvae	48 hrs.					56 hrs.					64 hrs.				
2 hrs.	94	86	86	80	N.S.	90	78	82	78	N.S.	86	76	82	74	N.S.
1 day	86	74	82	82	N.S.	82	68	74	78	N.S.	82	66	72	78	N.S.
2 days	90	82	90	84	N.S.	88	80	84	78	N.S.	84	80	82	74	N.S.
3 days	84	88	80	76	N.S.	82	84	78	76	N.S.	82	82	74	76	N.S.
4 days	90	94	92	94	N.S.	90	94	90	92	N.S.	90	90	86	88	N.S.
Larvae	72 hrs.					80 hrs.					88 hrs.				
2 hrs.	86	72	80	74	N.S.	84	72	80	74	N.S.	84	70	80	74	N.S.
1 day	82	66	72	74	N.S.	80	66	68	74	N.S.	78	66	66	74	N.S.
2 days	84	78	70	72	N.S.	84	78	80	72	N.S.	84	78	80	72	N.S.
3 days	80	80	70	76	N.S.	80	80	70	74	N.S.	80	80	70	74	N.S.
4 days	90	88	84	86	N.S.	88	84	82	82	N.S.	84	84	80	80	N.S.
Larvae	96 hrs.														
2 hrs.	82	70	80	74	N.S.										
1 day	78	66	66	74	N.S.										
2 days	84	74	80	72	N.S.										
3 days	80	80	70	74	N.S.										
4 days	82	82	76	76	N.S.										

Table 9. Results of exposure of striped bass egg and larval stages to less than atmospheric pressure (2 psia), followed by a return to atmospheric pressure, and subsequent exposure to the pumping cycle in the chamber shown in Figure 7, 8, and 9.

	Immediate			Day 1			Day 2			Day 3			Day 4		
Stage	Cntr. % al	Expt. % al	Sig.	Cntr. % al	Expt. % al	Sig.	Cntr. % al	Expt. % al	Sig.	Cntr. % al	Expt. % al	Sig.	Cntr. % al	Expt. % al	Sig.
Egg															
5 hrs.	100	100	N.S.												
11 hrs.	100	100	N.S.	54	54	N.S.									
13 hrs.	100	100	N.S.	88	86	N.S.	80	84	N.S.	70	58	N.S.	70	58	N.S.
21 hrs.	100	100	N.S.												
38.5 hrs.	100	100	N.S.												
42.5 hrs.	100	100	N.S.	100	100	N.S.	100	100	N.S.	100	100	N.S.	64	56	N.S.
44 hrs.	100	100	N.S.	92	100	N.S.									
48 hrs.	100	100	N.S.	100	100	N.S.	100	100	N.S.	100	100	N.S.	96	38	*
Larvae															
4 hrs.	100	100	N.S.	100	100	N.S.	68	34	N.S.	58	6	*			
10.5 hrs.	100	100	N.S.	100	100	N.S.	100	100	N.S.	100	100	N.S.	50	14	*
1 day 6 hrs.	100	100	N.S.	100	100	N.S.	100	100	N.S.	58	60	N.S.			
1 day 10.5 hrs.	100	100	N.S.	100	100	N.S.	100	100	N.S.						
2 days 13.5 hrs.	100	100	N.S.	100	100	N.S.									
2 days 20.5 hrs.	100	100	N.S.	100	100	N.S.									
2 days 22 hrs.	100	100	N.S.	86	84	N.S.									
3 days	100	100	N.S.	100	66	*	72	6	*						
3 days 4,5 hrs.	100	100	N.S.	92	70	*									
3 days 6 hrs.	100	100	N.S.	52	40	N.S.									
3 days 8 hrs.	100	100	N.S.												
4 days 8 hrs.	100	100	N.S.												
5 days	100	100	N.S.	100	100	N.S.	94	100	N.S.						
6 days 6.5 hrs.	100	100	N.S.												
8 days	100	92	N.S.	92	56	*	82	48	*	66	30	*			
11 days	100	100	N.S.	94	64	*	88	52	*	82	46	*			
12 days	100	100	N.S.	92	62	*	82	50	*	58	38	N.S.			
16 days	100	88	*	90	76	N.S.	84	64	*	52	44	N.S.			
17 days	100	100	N.S.	92	92	N.S.	82	82	N.S.	78	64	N.S.			

11.1 mg/l, ranging from 10.8 mg/l to 11.4 mg/l with a mean final dissolved oxygen value of 13.8 mg/l, ranging from 10.8 mg/l to 16.8 mg/l.

Based on the results of these experiments there would appear to be four stages which might be sensitive to this experimental regime. These stages are the very late egg stages, exposed just prior to hatch, the yolk-sac larvae exposed immediately after hatch, 3 to 4 day old yolk-sac larvae, and 8 to 16 day old larvae.

The results presented in Table 10 are based on three different experimental regimes. The exact experimental regimes are given in the materials and methods section. In both the experimental regimes used for the experiments reported in Part I and Part II of this table the organisms experienced the same rapid reduction to 2 psia from atmospheric pressure. These results support, except for the 24 day old larvae, the results presented in Table 9. The results presented for 24 day old larvae, in which the survival of the control group was less than that observed for the experimental group can be explained on the basis of chance and should not be considered to be representative of the survival of this life history stage to the pressure regimes experienced.

The three stages tested in Part I of Table 10 were exposed simultaneously to the same pressure regimes and conditions. Temperature and dissolved oxygen values remained constant at 17.9°C and 11.0 mg/l respectively during the entire exposure time for experimental and control groups. Experiments in Part II of Table 10 were run under the same conditions. The initial and final temperature measurements for control and experimental groups remained constant at 22.0°C and 21.5°C respectively. The initial dissolved oxygen values for the control and experimental groups remained constant at 10.2 mg/l. The final dissolved oxygen value for the experimental group was 10.4 mg/l, and for the control group 8.6 mg/l.

The five stages tested in Part III of Table 10 were exposed simultaneously to the same pressure regimes and conditions. The initial temperature values for the control and experimental groups remained constant at 20.4°C during the exposure. The final temperature for the experimental group increased to 21.5°C; for the control group it increased to 21.0°C. The initial and final dissolved oxygen values were the same for the experimental and control groups, 10.4 mg/l. The final dissolved oxygen value for the experimental groups decreased to 9.2 mg/l; for the control groups it decreased to 9.4 mg/l.

The results presented in Part III of Table 10 are based on the acclimation of the life history stages shown to 30 psig for 3 days,

Table 10. Results of exposure of striped bass egg and larval stages to three pressure regimes in the chamber shown in Figures 7, 8, and 9.

	Part I											
	Immediate			1 day			2 days			3 days		
	C	E		C	E		C	E		C	E	
Stage	% al	% al	Sig.	% al	% al	Sig.	% al	% al	Sig.	% al	% al	Sig.
Egg												
8 hrs.	100	100	N.S.									
45 hrs.	100	100	N.S.	96	96	N.S.	62	40	*	60	40	*
Larvae												
3 days 9 hrs.	100	98	N.S.	80	26	*	Experiment terminated					

	Part II					
	Immediate			1 day		
Larvae						
15 days	70	0	*	64	0	*
18 days	86	6	*	62	6	*
24 days	60	88	**			

	Part III[1]					
	Immediate			1 day		
Larvae						
13 days	40	2	*			
16 days	34	0	*			
16 days	64	0	*	58	0	*
17 days	78	28	*	58	20	*
22 days	0	10	N.S.	0	10	N.S.

[1]All experimental larvae that were dead, had died some time prior to release of pressure.

followed by a rapid reduction to atmospheric pressure. While the control survival in these experiments was not good, significant differences between control and experimental survival were observed immediately after exposure for all of the stages tested except the 22 day old larvae. It should be noted that the controls in this experiment were held in a 227 liter tank (the same volume as the pressure chamber) with no air-water interface for the same time period as the experimentals. While the results indicate significantly higher control survival for 13 to 17 day old larvae, it should be pointed out that upon removal from the test chamber, all the larvae that were dead had died some time before the pressure was released.

The results reported in Table 11 are based on the exposure of the macroinvertebrates indicated to a variety of pressure regimes. The mean initial temperature for the experimental groups was 20.8°C, with a range of 17.7°C to 25.9°C, and the mean initial temperature for the control groups was 21.1°C, with a range of 17.8°C to 26.7°C. For both groups the mean final temperature after exposure was 20.8°C, with a range of 18.0°C to 26.4°C. The mean initial dissolved oxygen value for the experimental group was 10.8 mg/l, with a range of 9.8 mg/l to 12.1 mg/l, and the mean final dissolved oxygen for the experimentals was 10.6 mg/l, with a range of 9.8 mg/l to 11.7 mg/l. For the control groups the mean initial dissolved oxygen was 10.7 mg/l, with a range of 10.0 mg/l to 12.1 mg/l, and the mean final dissolved oxygen value was 10.4 mg/l, with a range of 8.6 mg/l to 11.7 mg/l.

Gammarus species displayed variations in rates of mortality as a response to pressure exposures. Statistical analyses by contingency tables showed no significant differences between control and experimental groups under the following exposure programs: (1) pumping cycle alone; (2) negative pressure (2.5 to 3.4 psia) followed by the pumping cycle; (3) the generating cycle followed by negative pressure; (4) negative pressure, the pumping cycle, generating cycle, and then negative pressure. The fourth exposure program did show a consistently (although not statistically significant) higher mortality in the experimental group. When preceded by a 24 hour acclimation to 30 psig, the mortality was significantly higher in the experimental group for all observations (Table 11). *Gammarus* species subjected to negative pressure for five minutes followed by exposure to the pumping cycle showed a significantly higher mortality for the immediate observation compared to the control group (9 percent for the experimentals as opposed to 0 percent for the controls).

Monoculodes edwardsi and *Neomysis americana* showed no significant difference in viability from their control groups when

Table 11. Results of exposure of several macroinvertebrate species to a variety of pressure regimes in the chamber shown in Figures 7, 8, and 9.

		Immediate			1 day			2 days			3 days			4 days		
Organism	Treatment	C % al	E % al	Sig.	C % al	E % al	Sig.	C % al	E % al	Sig.	C % al	E % al	Sig.	C % al	E % al	Sig.
Gammarus sp.	NP	100	100	N.S.	98	96	N.S.	93	95	N.S.	85	88	N.S.	80	80	N.S.
Gammarus sp.	GN	100	100	N.S.	99	93	N.S.	90	85	N.S.	75	78	N.S.	68	70	N.S.
Gammarus sp.	NPGN	100	100	N.S.	94	90	N.S.	83	78	N.S.	74	71	N.S.	71	65	N.S.
Gammarus sp.	P	100	99	N.S.	91	88	N.S.	84	79	N.S.	76	75	N.S.	71	71	N.S.
Gammarus sp.	ANPGN	92	72	*	58	23	*									
Gammarus sp.	NP	100	91	*												
Neomysis americana	NPGN	51	67	N.S.												
Nonoculodes edwardsii	NPGN	98	100	N.S.	66	68	N.S.									
Chaoborus sp.	NPGN	100	96	N.S.	81	79	N.S.	74	72	N.S.	70	63	N.S.	67	57	N.S.
Chaoborus sp.	ANPGN	78	95	*	72	87	N.S.	68	78	N.S.	64	71	N.S.	64	58	N.S.
Polychaetes	NPGN	100	100	N.S.	100	100	N.S.	100	100	N.S.	100	100	N.S.	100	100	N.S.
Gastropods	NPGN	100	100	N.S.	100	100	N.S.	100	100	N.S.	100	100	N.S.	100	100	N.S.

A = Acclimation
N = Negative pressure
P = Pumping cycle
G = Generating cycle

subjected to negative pressure, pumping cycle, generating cycle, and then negative pressure.

Experiments on *Chaoborus* species, listed in Table 11, show that there were no statistically significant differences in mortality between control and experimental groups when exposed to negative pressure, pumping cycle, generating cycle, and negative pressure. In a subsequent experiment, with a full pressure program preceded by acclimation to 30 psig for 24 hours, the immediate observation showed a higher mortality in the control group.

The dipteran, *Chaoborus* species, displayed a complicated response to exposure to a pressure program consisting of acclimation for 24 hours to 30 psig, negative pressure, pumping cycle, generating cycle, and then negative pressure. The organisms utilized were in the larval stage of development, and the experiment was terminated when all organisms had either died or emerged as adults. Figure 6 shows that 83 percent of the control group metamorphosed into pupae as compared to 60 percent of the experimentals; 38 percent of the controls emerged as adults while none of the experimental organisms attained adulthood. Using survival analysis, there was no significant difference in death rates between control larvae and experimental larvae, but experimental pupae had a significantly higher death rate than the control pupae. For both groups the death rate of pupal organisms was higher than the death rates of their respective larval stages.

Hudson River gastropods and polychaetes showed 100 percent survival to 4 days after exposure to a pressure program of negative pressure, pumping cycle, generating cycle, and then negative pressure.

SUMMARY

The effect of pressure on entrained organisms depends on many factors. Changes in hydrostatic pressure during entrainment represent only one of several potential stresses experienced. The organisms will also be exposed to rapid changes in velocity shear forces, mechanical buffeting and turbulence.

The analyses presented in this paper, as well as information presented in the literature, suggest that each life history stage of an entrained species is likely to show differential survival rates, and that even within a given life history stage, variable survival values might well be expected.

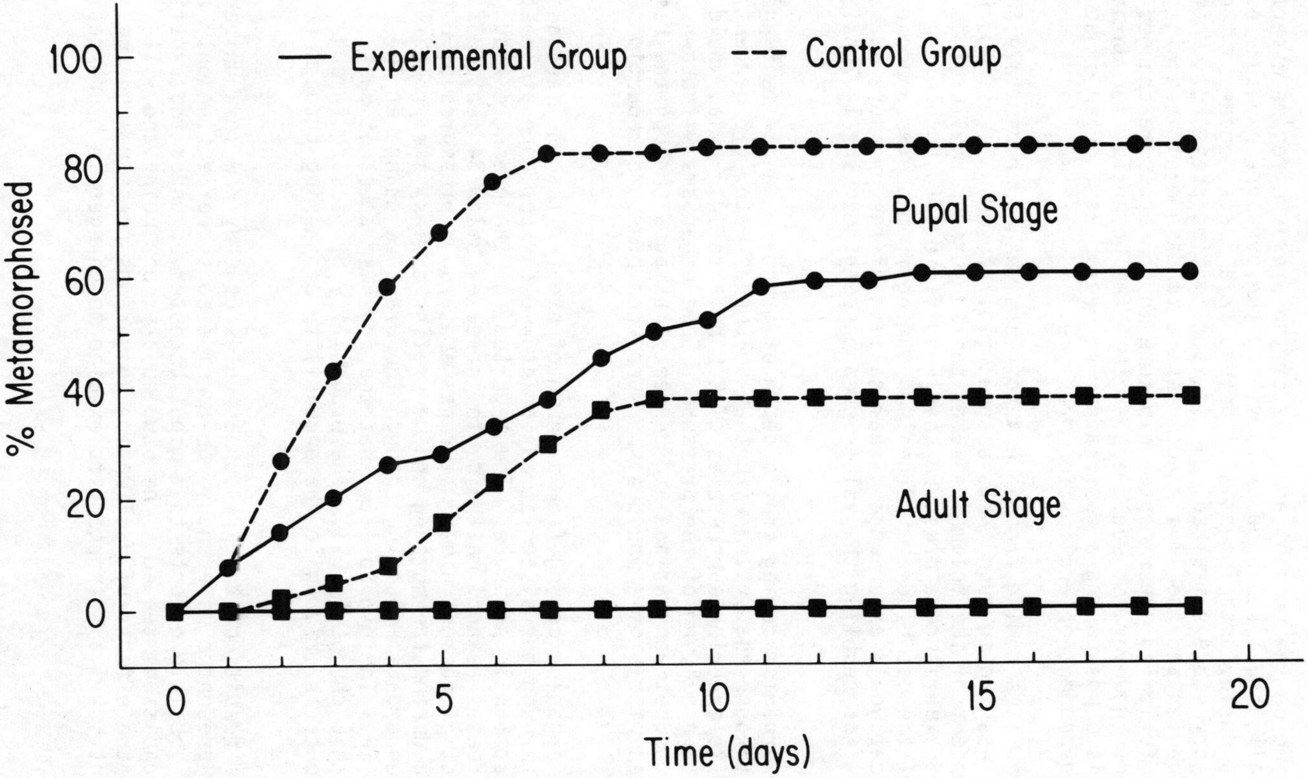

Figure 6. The percent Chaoborus species metamorphosed into pupae and adults.

Based on the current analysis of the pressure regime expected at Cornwall, the likelihood of surviving the pressure exposure experienced during entrainment will depend to a large degree on the life history stage entrained, and its acclimation pressure. Most of the species and the life stages tested to date exhibited high survival following experimental exposure to the pressure changes described. However, it would appear that the potential for damage to organisms possessing a gas bladder or hydrostatic organ of some type is dependent on acclimation pressure.

Organisms coming from a depth of less than approximately 34 feet and passing through the pressures (14.7 to 50.0 psia) incurred by 84 percent of the volume of water passed through the turbines would not experience a pressure change sufficient to double the volume of the gas bladder. Organisms coming from depths approaching 34 feet would experience a near halving of the pressure and a potential for doubling the volume when passing near the 14.7 psia region.

An organism coming from a depth of greater than 34 feet and passing through the system with that volume of water (16 percent) expected to experience pressures of 8.0 psia to 14.7 psia, would experience a halving of the pressure. Some organisms withdrawn from depths between surface and 34 feet that pass through the region associated with 8 psia would also experience pressure reductions of about half.

An organism coming from a depth of greater than 34 feet and passing through the system at any point would experience at least some reduction in pressure. The magnitude of the pressure reduction increases as acclimation depth increases below 34 feet. If 16 percent of the entrained water passes through pressures of 8 to 14.7 psia during the pumping and generating cycles, the probability of a given organism passing through the pressure change at least once is .29. The probability that the organism does not pass through this change is .71. The probability of passing through exactly once is .27, while the probability of passing through twice is .03.

The results of the pressure studies done by New York University and presented here should be considered as reflecting only the range of possible effects that might be observed under the experimental conditions tested. The studies are continuing and will focus on the effects of acclimation to various pressures on the subsequent survival of organisms exposed to changes in hydrostatic pressure.

LITERATURE CITED

BISHAI, H. M. 1961. The effect of pressure on the survival and distribution of larval and young fish. J. Conseil, Conseil Perm. Intern. Exploration Mer. 26: 292-311.

DAMANT, G. C. C. 1924. The adjustment of the buoyancy of the larva of Corethra plumicornis. J. Physiol. Lond. 59: 345-356.

DIGBY, P. S. B. 1972. Detection of small changes in hydrostatic pressure by Crustacea and its relation to electrode action in the cuticle. In M. A. Sleigh and A. G. MacDonald (eds.), The Effects of Pressure on Organisms, Academic Press, Inc. New York, New York, pp. 445-471.

ENRIGHT, J. T. 1961. Pressure sensitivity of an amphipod. Science 133: 758-760.

ENRIGHT, J. T. 1962. Responses of an amphipod to pressure changes. Comp. Biochem. Physiol. 7: 131-145.

EBBECKE, U. 1935. Uber die Wirkungen hoher Drucke auf Marine Lebewesen. Arch. Ges. Physiol. 236: 647. In A. M. Zimmerman (ed.), High pressure effects of cellular processes, Academic Press, New York and London (1970).

EBBECKE, U. 1944. Lebensvorgange unter der Einwirkung hoher Drucke. Erbeg. Physiol. 45: 34-185. In O. Kinne (ed.), Marine Ecology, Vol. 1, John Wiley and Sons Ltd. (1972).

KINNE, O. (Ed.). 1972. Marine Ecology, Vol. I, Part 3. John Wiley and Sons, Ltd., New York.

KNIGHT-JONES, E. W. and E. MORGAN. 1966. Responses of marine animals to changes in hydrostatic pressure. Oceanog. Marine Biol. Ann. Rev. 4: 267-299.

LAGLER, K. F., J. E. BARDACH and R. P. MILLER. 1962. Ichthyology. John Wiley and Sons, Inc.

NAROSKA, V. 1968. Vergleichende Untersuchungen über den Einfluss des hydrostatischen Druckes auf Überlebensfahigkeit und Stoffwechselintensität mariner. Everbraten and Teleosteer. Kiel Meeresforsch. 24: 95-123.

SLEIGH, M. A. and A. G. MacDONALD (eds.). 1972. The effects of pressure on organisms. Symposia of the Society of Experimental Biology. Academic Press, New York.

TSVETKOV, V. I., D. S. PAVLOV and V. K. NEZDOLIY. 1972. Changes of hydrostatic pressure lethal to the young of some freshwater fish. J. of Ichthyology 12: 307-318.

ZIMMERMAN, A. M. (Ed.). 1970. High pressure effects on cellular processes. Academic Press, New York.

THE EFFECTS OF TWO ELECTRICAL BARRIERS ON THE ENTRAINMENT OF FISH AT A FRESHWATER NUCLEAR POWER PLANT

Martin A. M. Hyman, William H. Mowbray
and Saul B. Saila
Marine Experiment Station
University of Rhode Island, Kingston, Rhode Island 02881

ABSTRACT

In order to limit the numbers of fish entrained at the Connecticut Yankee Atomic Power Plant (CYAP) at Haddam Neck, Connecticut, it was desirable that a system be designed to discourage the presence of fish from the immediate environs of the plant intakes. Two such systems are described, both of which statistically reduced the numbers of fish taken. The feasibility of electrical barriers for reducing power plant entrainment in a freshwater environment was demonstrated.

INTRODUCTION

The Connecticut Yankee Atomic Power Company plant is located on 550 acres of land at Haddam Neck on the Connecticut River near Middletown, Connecticut, about 22 km. from the river mouth. The power plant commenced commercial operation on January 1, 1968, and produces approximately 600 megawatts of electricity. The reactor is a pressurized, light water-cooled, water moderated type with a thermal power rating of 1825 megawatts, one of the largest plants of its kind in the world. The plant utilizes approximately 880 cubic feet per second (25 m^3/sec.) of cooling and service water from the river in a once-through system. The discharge water is heated a maximum of 12.5°C above ambient before it is returned to the river via a canal approximately 1.8 km. downstream from the intakes. The intake screenwell structure is located on the east bank of the river and contains four intake bays. Each bay is 3.77m. wide and 5.5m. deep below mean low water. Under normal conditions intake velocity is less than 30 cm/sec. The plant is located about 4.8 km. above the salt wedge, thus avoiding the additional harsh effects of the marine environment. However, the river in this area is still under tidal influence, with an average rise and fall of 0.75m. Although the predominant flow of the river is downstream, reverse flow does occur on flood tides except during the spring freshets.

As part of the licensing requirement, the Connecticut River Ecological Study (Merriman, 1965-1972) was initiated in 1965 to survey the impact of the plant. In addition to sponsoring this seven year study, the Connecticut Yankee Atomic Power Company and the Northeast Utilities Service Company have also sponsored studies designed to minimize the impingment of fish by the plant's cooling water intakes. Since the Connecticut Yankee plant began operating, there have been several isolated occurrences of relatively large numbers of fish being taken on the travelling screens of the plant intakes. In each case the fish involved were anadromous species of the family Clupeidae. Preliminary investigations to determine the cause of such occurrences were initiated in 1967 and included identification and vital statistics of fish taken on the screens, along with collection of environmental data with the intent of correlating the occurrence of impingement with environmental variables. The results were inconclusive.

In June 1970, a study was initiated by Northeast Utilities Service Company to determine the effectiveness of an electric fish barrier located in one of the four intake bays. A comparison was made with the average numbers of fish taken in the other three bays which had no electrical barriers. A description of the barrier, the control unit and some results were provided in Connecticut River Ecological Study, Twelfth Semi-Annual Report. This study indicated that although a reduction in fish was found in the single bay equipped with the electrical barrier, the experimental results were not dependent upon the energized state of the electrodes and that the repelling effect may have been associated with sound pressure fields created by the aluminum conduit electrodes or possibly their presence as a visual cue. It was obviously difficult to make inferences from this preliminary experiment which involved only one barrier and no effective controls; consequently more detailed studies were planned.

Review of past years' impingement records has since shown extremely high variability among the records of the four bays. This may be the result of random, partial clogging of the screens with resulting change in pressure, or of blocking of the outside trash racks with a resulting decrease in cross-sectional area and a localized increase in velocity. Increased velocity may produce a higher probability of impingement, or the clogging of the outer racks may permit an avoidance. Moreover this blockage may present itself as a shelter and have the effect of concentrating animals in the area. Data are not available to permit separation of the above factors which affect fish behavior.

Following the completion of the single bay study in 1970, identical barriers consisting of aluminum conduit arrays were installed in the remaining three bays. Ultimately, individual electric controllers for each barrier were also installed.

Additional experiments were then conducted with the projected goal of minimizing the number of fish captured on the plant intake screens. The specific intent of this study was to compare objectively the relative effectiveness of the existing barrier array when it was electrically energized, with the barrier array operating without electrical stimulation. Any significant results, regardless of magnitude, would be impetus for further studies. By no means was it projected that the first system designed would be the optimum for all species and all sizes. Prior to the 1972 evaluation of the Phase I electrode study, personnel of the University of Rhode Island's Marine Experiment Station had served only as consultants, advising that an electrode array in only one bay was insufficient. On the basis of the positive evaluation of Phase I the Marine Experiment Station was asked to design, fabricate, install and evaluate a Phase II design. By the spring of 1973 the new system was awaiting evaluation, but flood conditions and plant operational difficulties resulted in delay of the survey. Fall refueling caused further delay until an evaluation was no longer advisable because of low fall levels of fish abundance. The evaluation was postponed until the summer of 1974.

BACKGROUND

Many systems have been applied to alter the behavior of animals. Recent technological advances include those of Ellis and Hoopes (1972) and Johnson and McNeeley (1966). Maximizing yields while minimizing effort has been an important result of new methods. However, another facet of response is often equally important. For preservation, management, and protection, it is sometimes beneficial to restrict an organism's access to a certain area; thus, the need for man-made biological barriers is widespread (Erkkila <u>et al</u>., 1956; Smith, 1966). Frequently it is desirable that these barriers be selective, as is often the case in nature. Many organisms are limited by reasonably distinct boundaries of such variables as temperature, pressure, and salinity (Moore, 1958).

If the list of possible barrier systems were enumerated, it would be extremely brief. Without question, the number of physiological systems available for stimulation are limited. Of these systems, many have thresholds which are impractical to work with or have not been sufficiently well demonstrated (i.e., sensitivity to magnetic fields). Others, such as olfaction and gustation, have limited effectiveness. The system described here is an electrical barrier, applied at a fresh water power plant intake.

The effects of electricity have been used by the Russians to increase the efficiency of fish pumps (Woodhead, 1963; Nikonorov,

1964), and it has been reported (Johnson and McNeely, 1966) that by 1963, 80 United States east coast menhaden purse seiners were already using electric field fish pumps. There has also been a great deal of effort to use electrical effects to increase fish catches in pelagic fish trawls (Anonymous, 1965), and in the United States to develop electric shrimp and lobster trawls (Pease and Seidel, 1967; Saila and Williams, 1972). The general problem of the application of electricity to inland fishery biology and management has been considered in a symposium edited by Vibert (1967), and Meyer-Waarden, Halsband and Halsband (1960) provide a good bibliography on electrical fishing both in fresh water and in the sea.

In spite of the fact that a patent for an electric fish barrier was granted in the United States some time ago (Burkey, 1917), relatively little has been done to improve systematically the performance of such devices. Electrical screens or barriers designed to stop or divert upstream migrant fish of large size are comparatively easy to design and have proven effective both in the United States (Burrows, 1957) and in Europe (Lethlean, 1953). However, it is very difficult to design electrical barriers for sites where fish follow the flow of water, including downstream movement in streams and the intakes of power plants. In these cases, any significant interference with the ability of a fish to swim causes it to be washed further into the area to be protected and thus decreases the probability of escape. Pulsed currents are desirable in these situations to avoid interfering with the swimming ability of the fish. Ideally, the barrier should be very carefully located where it can be used to provide a safe by-pass away from the attraction of the intake. The basis of a successful electrical barrier is that the fish must be introduced gradually into an unpleasant sensation, allowed to learn to avoid it, and then be persuaded to go elsewhere.

In the presence of an electric field, aquatic animals demonstrate several distinct behavioral patterns depending on the species, the size of the animals, and the field characteristics. The latter are a function of the electrode configuration and the environmental variables, such as conductivity and temperature. The various behavioral responses have been demonstrated on a fairly wide range of animals. On approaching a field many animals will exhibit a fright reaction and be repelled. This response is the key to a barrier system. Depending on the shape of the field, as a fish proceeds deeper into the field it will be subjected to a greater field strength. When this exceeds a "threshold" level, normal swimming behavior is inhibited. In a direct current (DC) field, a fish often exhibits an electrotactic response of stimulated swimming toward the anode. The larger the fish the more voltage it is subjected to, and the greater the potential difference over its length. This statement must, however, be qualified. If the animal is aligned by chance or by design along a line of equipotential, then voltage is effectively eliminated across its body length and the fish is subjected only to a lower voltage across, and proportional to its width. Consequently, depending upon alignment, the longer the fish the lower the required field intensity required. The resulting length-voltage relationship is inverse and non-linear (Leathlin, 1953). If a fish is placed in a uniform field which is large in extent compared to the size of the fish, the current density and the voltage gradient across it will be determined by the ratio of the resistivity of the water to that of the fish. If this ratio is greater than unity, as would normally be the case for fresh water, the current density in the fish would be larger, and the voltage drop across the fish would be smaller than that in the surrounding medium. The fish would be a better conductor than the water around it.

EQUIPMENT AND METHODS

PHASE I (1972)

Each of the four intake bays (labeled A, B, C, D from north to south) was equipped with an underwater electrode array connected to a "Burkey Fish Fence Charger" furnished by the Electric Fish Screen Company, Hollywood, California.* Model 2A-240 chargers were used on bay B and bay D. This arrangement was chosen to permit one of each type of charger on an inside and an outside bay. Both charger types employ a capacitor discharge system controlled by an ignitron and appropriate timing circuits to produce rapid rise, short duration, high voltage pulses. Figure 1 illustrates the electrode configuration and orientation with respect to the intake. The four electrode arrays were identical and consisted of 15 vertical aluminum pipes: 6.1 m. lengths of 2.5 cm. nominal i.d. conduit spaced 19 cm. on center and connected with alternate positive and negative polarities, the first and last electrode being positive. Figure 2 illustrates the wiring of the electrodes. It should be noted that there were two additional arrays of aluminum pipes located 46 and 92 cm., respectively, behind the first array which were identical to the first but not electrically connected. Although generally considered a poor electrode material due to corrosion and changes in its electrical characteristics,

---

*Use of trade names does not imply endorsement of the manufacturers or their products by the authors or the University of Rhode Island.

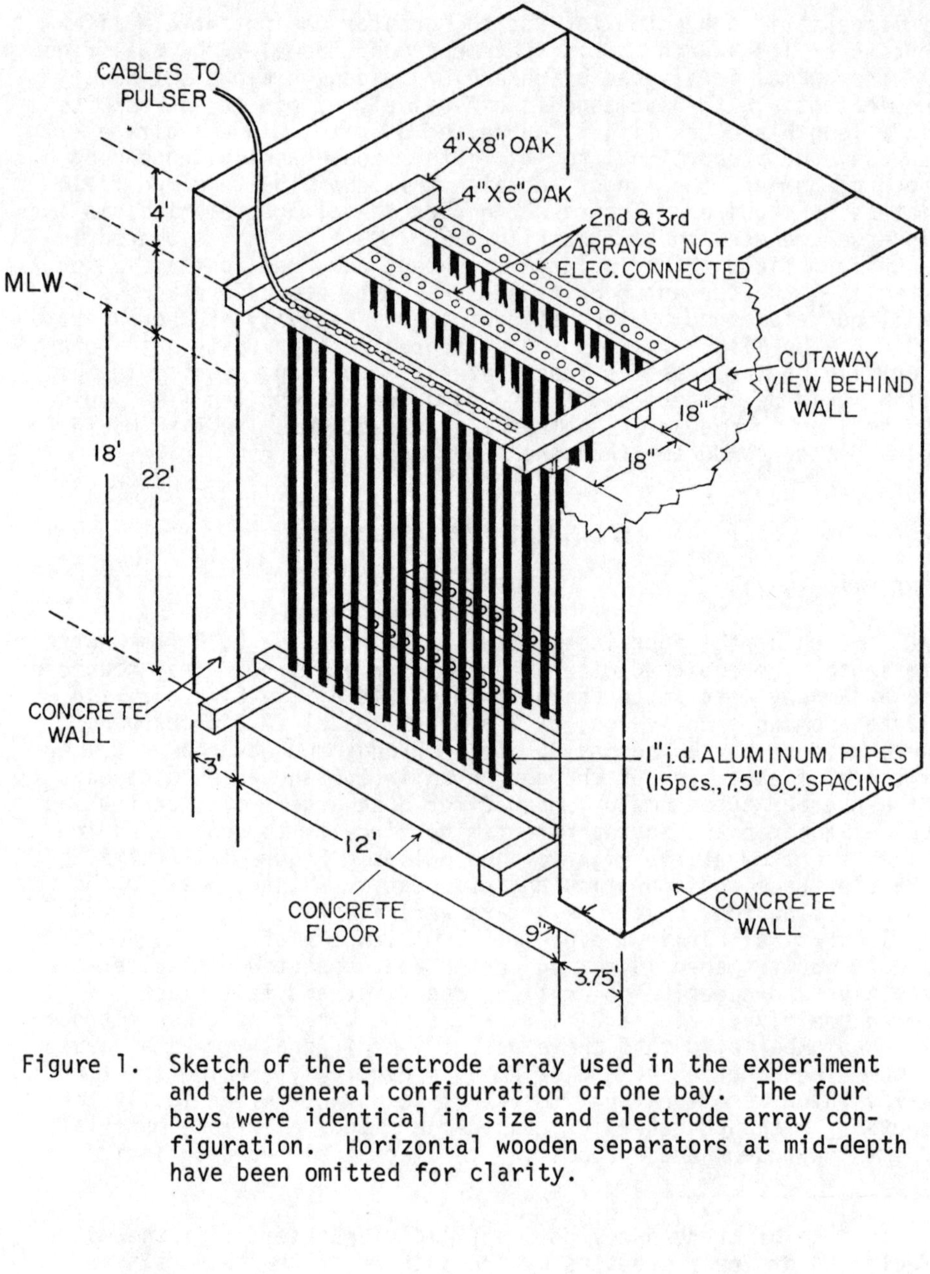

Figure 1. Sketch of the electrode array used in the experiment and the general configuration of one bay. The four bays were identical in size and electrode array configuration. Horizontal wooden separators at mid-depth have been omitted for clarity.

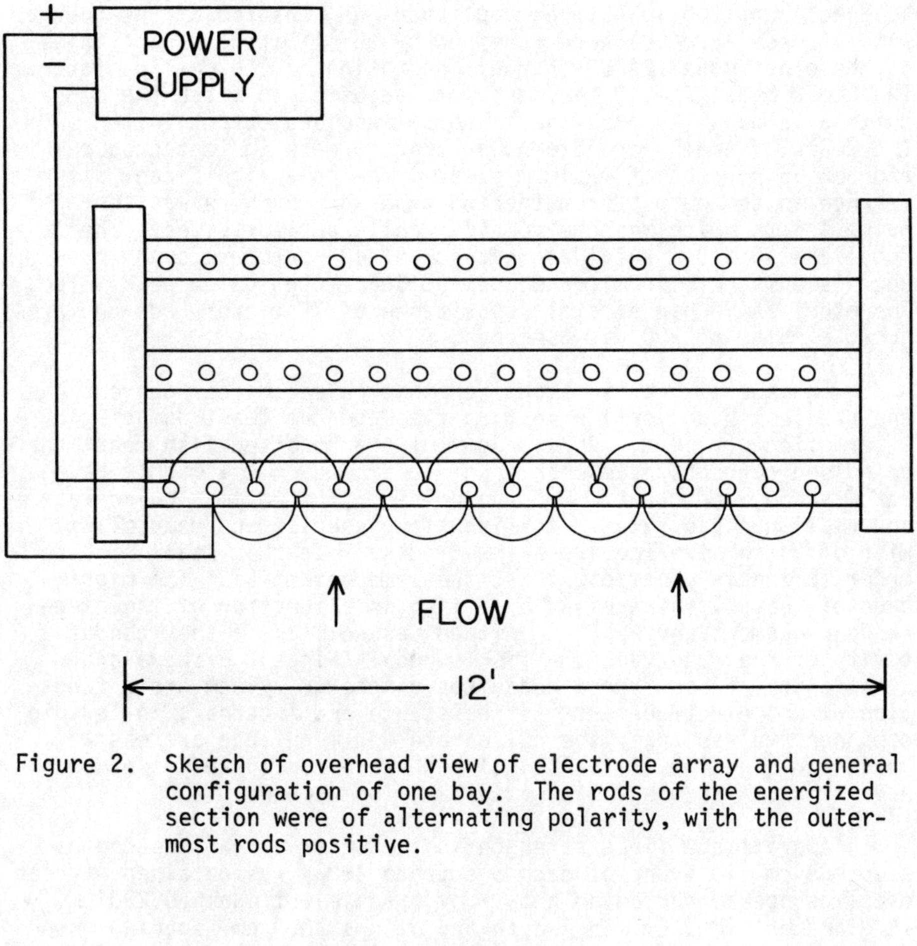

Figure 2. Sketch of overhead view of electrode array and general configuration of one bay. The rods of the energized section were of alternating polarity, with the outermost rods positive.

aluminum was chosen because it was inexpensive, available and light. At no time were these experimental setups considered permanent or optimum.

Initial conditions of 144 pulses per minute (approximately 2.5/sec.) and 700 volts peak amplitude (as measured at the charger with an oscilloscope) were governed by the limiting capabilities of the older Model 2A-50 chargers operating at 100 percent output. The newer Model 2A-240 chargers were adjusted to match the older units as closely as possible in pulse rate and output voltage. The dials of their variable-power transformers indicated operation at 56 percent of maximum power. The only significant difference in the output from the two types of chargers was the pulse length which was due to different capacitor sizes. The Model 2A-240 had a total capacitance of 40 microfarads and produced a pulse of 150 microseconds at 50 percent below peak value. The Model 2A-50 had a total capacitance of 75 microfarads and produced a pulse of 300 microseconds.

In general, it is recognized that pulsed direct current has the greatest behavioral effect on fish and the least damaging after-effect (Vibert, 1967). Some of the reported fish reactions to pulsed currents have been found to depend on the magnitude, rate, shape, and duration of the pulses. The optimum pulse rate and amplitude are related to fish size, species and metabolism. Fish of large size require a lower pulse rate than small fish in order to induce electrotaxis at the same potential. The magnitude of the pulses received by a fish is a function of the electrode placement and applied voltage, as well as of the conductivity of the water and the fish's body. Since the shape and duration of pulses from a capacitor discharge system are a function of the electrode array's resistance and internal, inflexible component values, only the pulse rate and magnitude are easily variable. The initial values of both were considered a reasonable approximation on the basis of information at hand.

Approximate field strength in the water at various depths about 38 cm. in front of each electrode array was obtained with an oscilloscope connected to a pair of brass electrodes (0.238 cm. in diameter, 10.0 cm. in length and with a 10.0 cm. spacing) supported on a fiberglass pole. Peak readings varied from 0.26 to 0.31 v/cm. with the sampling electrodes parallel to the arrays and were approximately proportional to their respective electrode currents. Potentials varied between 0.03 and 0.06 v/cm. with the sampling electrodes perpendicular to the arrays. Thus a typical 20 cm. total length fish approaching the system from the normal would be subjected to a 0.60 to 1.20 volt differential over his body length. A typical 1 cm. wide fish would be subjected to a

0.26 to 0.31 volt differential over this dimension. The average resistivity of the water in this region was approximately 1000 ohm-cm.

The experimental design utilized for comparing the relative effectiveness of the energized versus unenergized electrode array is called the double change-over design. Federer (1955) illustrates the applications of this design and indicates its advantages. They include:

a) The residual effects as well as direct effects of treatments can be estimated.
b) A high degree of precision is usually attained.
c) The design is suitable for small numbers of treatments.

The linear model for the gth repetition of the double change-over design is:

$$X_{ijhg} = \mu + \alpha_i + \beta_{jg} + (\alpha\beta)_{ijg} + \delta_h + P_p + \varepsilon_{ijhg},$$

where μ = mean effect, α_i = effect of ith period, β_{jg} = effect of jth sequence in the gth repetition in the ith period, $(\alpha\beta)_{ijg}$ = effect of the jth sequence in the gth repetition in the ith period, δ_h = direct effect of the treatment, P_p = residual effect of the preceding treatment, and ε_{ijhg} = random component of error; g = 1, 2, ..., q; i = 1, 2, ..., k; j = 1, 2, 111, k (k-1); h = 1, 2, ..., k; and p = 1, 2, ..., k.

Size and species of all organisms captured were determined at six hour intervals for the duration of the experiment. This capture process involved regular washing of the travelling screens individually at six hour intervals. The fish were washed into the power plant trash basket which was equipped with a 1/2-inch mesh liner of hardware cloth. All organisms were identified to species, measured and weighed. All fish were measured to the nearest millimeter (total length) and weighed to the nearest two grams. Unknown or questionable specimens were preserved for later identification.

PHASE II

At no time was the electrical barrier used in Phase I considered optimum in terms of physical location or electrical characteristics; it was employed only to perform initial tests on the existing equipment. Following the encouraging results of Phase I, improvement of various aspects of the system was indicated. It was felt that the vertical configuration and location of the Phase I electrodes were faulty on several points. A fish approaching the intake structure would be inside a bay before it could detect a field of any strength. Even in very low intake velocities, this would obviously be an adverse situation. Consequently, it would be beneficial to move the field away from the immediate vicinity of the intake in order to maximize its effect. Secondly, it would be beneficial to have the field encompass a greater expanse than did the Phase I system. Finally, it has been accepted that the optimum method of repelling fish is to have them come under the influence of the field gradually to make maximum utilization of the fright response. Taking all of these into consideration, the design of the Phase II system was conceived as illustrated in Figure 3. The four cathodes (-) are suspended from an oil slick boom and submerged approximately 0.92 m. Each section is an 8.2 m. length of 15.5 cm. diameter aluminum irrigation pipe. The boom is anchored to approximate a truncated equilateral triangle about 17 m. on each short side. The anodes (+) are located on the river bottom, running parallel to, and approximately 10.6 m. from the corresponding cathodes. Each section is 18.3 m. long and is made of three lengths of 30.5 cm. diameter aluminum drainage pipe connected by bolted straps. The depth is 1.5 to 5.5 m. at normal river level, varying with bottom topography. Each anode is paired with its parallel cathode; each electrode pair is powered by a single pulser. Subsequent survey found the bottom electrodes to be largely buried by silt and sand. However, this does not hamper system functioning since the conductivity of the sediment is at least equal to that of the overlying water column.

During the Phase II study the area in front of the intake was equipped with a new underwater electrode array powered by four Model 2A-240 "Burkey Fish Fence Chargers." The original electrode system inside the four bays used in Phase I was left intact, but not connected electrically. In order to monitor the system performance and detect any malfunctions, Pearson Electronics Model 110 pulse-type current transformers were installed in the negative output cable of each charger, the signals from which could be selected by a switch and observed on a Phillips PM3200 oscilloscope. During each of the "on" periods the output currents were monitored and recorded; abnormal values could be traced by equipment checks or electrode inspections allowing immediate remedial action to be taken.

During the Phase I study, the four pulsers fired asynchronously; however, in Phase II it was desirable to have all four units pulse simultaneously because of the overlapping electric fields. To that end, the triggering units of the four pulsers

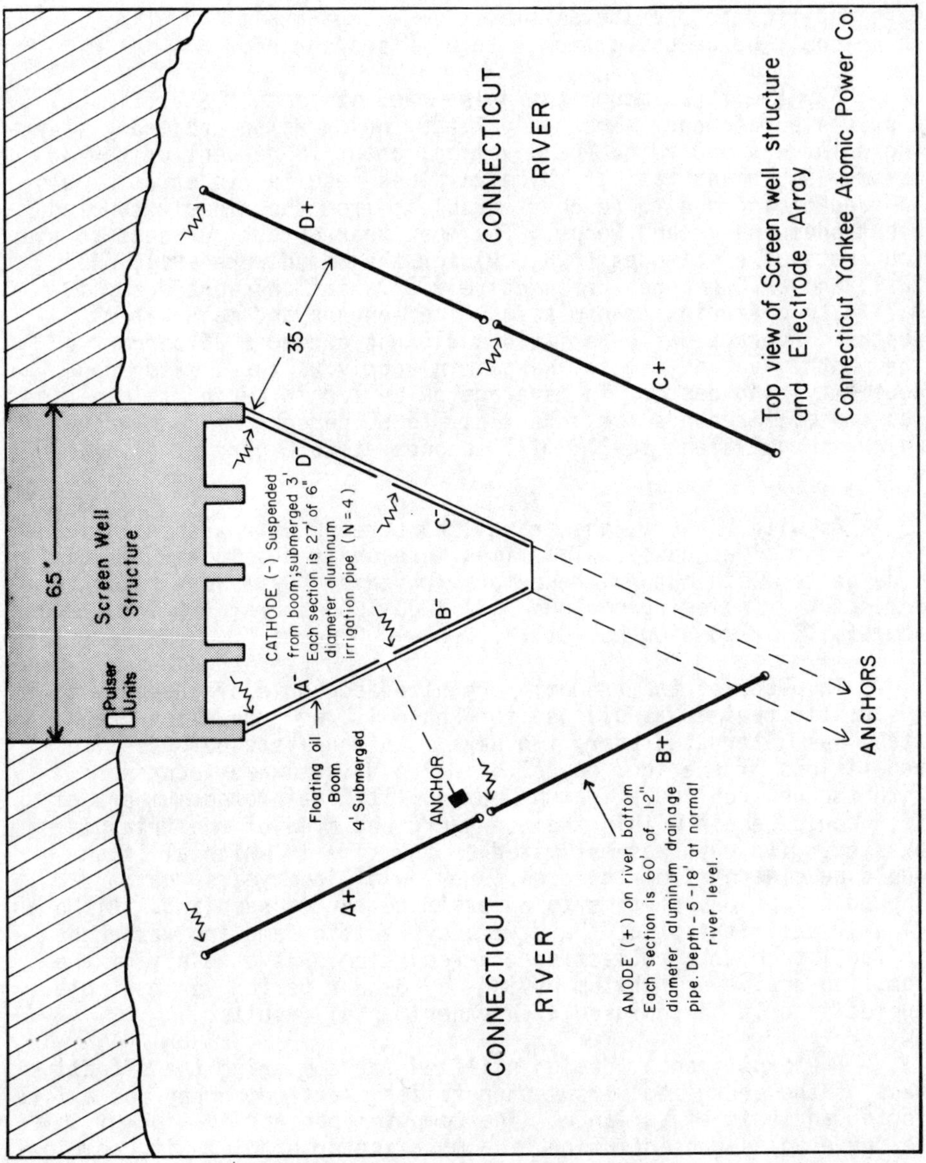

Figure 3. Top view of screen well structure and electrode array illustrating electrode configuration and anchoring system.

were modified to operate as slaves to a separate master timer which could be adjusted from 2 to 6 pulses/second.

For the first month the pulse rate was set at approximately 2 pulses per second. The d-c voltage chosen as an arbitrary starting point was 950 volts (peak value), about 70 percent of the maximum equipment rating. Higher values resulted in erratic pulsing and timing due to feedback problems from the closely coupled electrodes and ground loops. The mean peak output currents to the four sets of electrodes (A,B,C,D) for the month were 99.1, 66.7, 82.1, and 87.8 Amperes, respectively. Variations were due primarily to differing transmission line lengths and mean water depths. Average daily variations did not exceed ± 25 percent of the monthly values due to changes in supply voltage, water conductivity, and depth. The average pulse length during this period was 1.9 milliseconds for a decrease to 50 percent of peak value which is equivalent to 2.7 milliseconds at 36.8 percent (i.e., ε^{-1}).

As with Phase I, the travelling screens were washed individually four times daily. The times were shifted slightly to coincide better with natural behavioral rhythms. Data were taken to reflect the following periods: I) 1000-1600 hours; II) 1600-2200 hours; III) 2200-0400 hours; IV) 0400-1000 hours.

In place of the randomly organized schedule of Phase I, a systematic regime was planned for Phase II, and the electrode state was alternated every two days. For analysis purposes, seven repetitions of the four day (2 on - 2 off) set were recorded. With the new configuration the possibility that large numbers of fish could remain within the non-energized area of the triangle existed. Two days was estimated as the time in which all fish would be cleared from the area. Review of past years' data showed no obvious rhythms to necessitate random sampling. Although clearly a limitation of the study, systematic sampling was chosen to facilitate data collection and reduction. As a result of the sampling regime any rhythm having a 48-hour period (or multiple thereof) could be confused with experimental results.

The experimental design utilized for comparing the effectiveness of the energized versus unenergized electrode array was a five factor analysis of variance. The computerized application of this design enables consideration of a multifactor problem with complex interaction among terms. The five factors considered here were Bay (A,B,C,D); Period (I, II, III, IV); Day (day 1 or 2 of treatment); Electrode State (on, off); Repetitions.

As before, size and species of all organisms collected were determined at six-hour intervals for the duration of the experiment. For this phase of the study, no 1/2-inch mesh liner was used in the collection basket. Consequently, smaller specimens may have been lost; it is assumed that this loss would be a random event with respect to electrode state. Furthermore, only extremely small specimens would be lost, most of which are of a size less than that sensitive to the electric field.

The results of the initial part of the Phase II survey suggested the system variables would have to be tuned to obtain optimum performance. The data suggest that two days may not have been a long enough "on" period. If the system were effective, it would be possible to have fish trapped with no way to escape the field other than toward the plant. Without any expeditious method of sweeping this area clean, or of estimating the residual population, the only recourse was to lengthen the "on" time and determine if there was any extinguishing or if the levels remained reasonably constant over time. In reviewing the electrode configuration, it was also found that if a fish entered the field and attempted to escape by diving to the bottom, it could effectively break through the barrier and continue toward the plant. To alleviate this situation chains were suspended from the negative electrodes. Along the length of the electrodes, 3.35 m. pieces of .48 cm. galvanized chain were placed at 61 cm. intervals. These chains reached the bottom over more than 50 percent of the array and effectively increased the size of the cathode and its respective field, thus better protecting the bottom areas. Because the first pulse rate may have been too low to stimulate the size fish present efficiently, the rate was increased from 2 to 6/second. The generally small fish available (as opposed to adult salmon) and the predicted fall run of juveniles, linked with the poor results of the June investigation were ample justification for the rate elevation. The decision to make several modifications was made in the interest of conserving time; however, it is recognized that without controlled experiments there is no way to determine which of a series of modifications is successful.

The survey was then continued for one month (September) at the previous voltage level and the higher pulse rate, and then for an additional month (October) at a decreased voltage level. Specifically, the input voltage was reduced by a factor of $\sqrt{2}$, thus reducing the power by 50 percent and the d-c peak electrode voltage to 650.

This two month period brought the survey to 1 November. In the interest of obtaining maximum information to aid in future decisions regarding barriers of this design, it was opted to continue the study with a series of mini-experiments. With no trends presenting themselves during the one week "on" series, it was considered safe to reduce the duration to three days. A series of three days off/three days on experiments was initiated, projected to continue until the population declined to levels which prohibited further study. Each on period was to be compared with the two flanking off periods. These experiments are in progress.

RESULTS

PHASE I

Preliminary examination of the list of species illustrated in Table 1 demonstrated that of the total of 24 species captured, only the catfishes, perch and some centrachid fish were sufficiently abundant to have appeared fairly consistently as relatively large specimens. The species captured were divided into two groups with reference to their ecological behavior. These were: a) all catfish species and eels, and b) all other species. However, since large fish are subject to a higher body voltage than smaller fish an arbitrary selection of only those fish greater than 10 centimeters in total length was made for analysis in the experiments to be described.

The data were collected at six-hour intervals, recording the fish from each bay separately. The state of the electrode for Period I was determined randomly each day, followed by alternation of the electrode states at each wash for the following three periods.

Table 2 describes the summary data for the four bays using the numbers of catfish species and eels as the response variable. The results of this preliminary analysis suggest that the average number captured with the electricity on was consistently smaller than the number captured with it turned off. However, these summary data give no indication of the statistical significance of this apparent increase in barrier efficiency with respect to electrode state.

To test whether there is a significant difference in the relative proportions of catfish and eels by time period and treatment, a Chi-square analysis was performed (Table 3). The Chi-square value of 13.12 for three degrees of freedom is statistically significant. The null hypothesis of independence in relative proportions by time periods and treatments is rejected at the $\alpha = 0.05$ level.

Table 1. List of fish species captured during the experimental period (Aug. 18 - Sept. 15, 1972) at Connecticut Yankee intake screens. (Alphabetically by genus.)

Scientific Name	Common Name	Size*	Abundance**
Alosa aestivalis	Blueback herring	S	C
Alosa pseudoharengus	Alewife	S	C
Alosa sapidissima	American shad	S, L	R
Anguilla rostrata	American eel	L	R
Brevoortia tyrannus	Atlantic Menhaden	S	C
Catostomus commersonii	Common sucker	L	R
Cyprinus carpio	Carp	L	R
Esox lucius	Northern pike	L	R
Esox niger	Chain pickerel	L	R
Etheostoma nigrum	Johnny darter	S	R
Ictalurus catus	White catfish	S, L	C
Ictalurus nebulosus	Brown bullhead	S, L	A
Ictalurus punctatus	Channel catfish	L	R
Lepomis auritus	Redbrest sunfish	S	R
Lepomis gibbosus	Common sunfish	S, L	C
Lepomis macrochirus	Bluegill	L	R
Menidia menidia	Silverside	S	R
Morone americanus	White perch	S, L	A
Notemigonus crysoleucas	Golden shiner	S	R
Notoropis cornutus	Common shiner	S	R
Notoropis hudsonius	Spottail shiner	S	C
Perca flavescens	Yellow perch	L	C
Pomoxis nigromaculatus	Black crappie	L	R
Trinectes maculatus	Hogchoker	S	R

*S refers to small - less than 10 cm. T.L.
L refers to fish greater than 10 cm. T.L.
**R = rare, C = common, A = abundant

Table 2. Summary data for the four bays showing numbers of catfish species and eels greater than 10 cm. total length captured by periods.

Bay A Data (24 days)

Period	Pulser ON	Pulser OFF	Sum
I (06:00-12:00)	11	11	22
II (12:00-18:00)	12	9	21
III (18:00-24:00)	3	7	10
IV (24:00-06:00)	10	16	26
Total	36	43	79
Mean	1.50	1.79	

Effect: 16.2 percent reduction with pulser ON.

Bay B Data (20 days)

Period	Pulser ON	Pulser OFF	Sum
I (06:00-12:00)	16	11	27
II (12:00-18:00)	8	34	42
III (18:00-24:00)	21	22	43
IV (24:00-06:00)	18	24	42
Total	63	91	154
Mean	3.15	4.55	

Effect: 30.8 percent reduction with pulser ON.

(cont'd.)

Bay C Data (24 days)

Period	Pulser ON	Pulser OFF	Sum
I (06:00-12:00)	22	7	29
II (12:00-18:00)	10	27	37
III (18:00-24:00)	11	14	25
IV (24:00-06:00)	11	20	31
Total	54	68	122
Mean	2.25	2.83	

Effect: 20.6 percent reduction with pulser ON.

Bay D Data (24 days)

Period	Pulser ON	Pulser OFF	Sum
I (06:00-12:00)	12	18	30
II (12:00-18:00)	15	13	28
III (18:00-24:00)	19	18	37
IV (24:00-06:00)	13	24	37
Total	59	73	132
Mean	2.46	3.04	

Effect: 19.1 percent reduction with pulser ON

Table 3. Numbers of catfish species and eels captured during the four periods classified according to whether the pulser was ON or OFF (combined data for the four bays). Lower case f denotes observed frequencies and capital letter F denotes expected values.

Period	Pulser ON	Pulser OFF	Total
I	f 61	f 47	108
	F 47.014	F 60.985	
II	f 45	f 83	128
	F 55.72	F 72.279	
III	f 54	f 61	115
	F 50.061	F 64.938	
IV	f 52	f 84	136
	F 59.2	F 76.796	
	212	275	487

$X^2 = 13.120$ d.f. = (R-1) (C-1) =(4-1) (2-1) = 3

$X^2_{.05}$, d.f.3, = 7.81

Since the fish were collected at six-hour intervals, this series of experiments involved four test periods per day. In such an experiment the total number of degrees of freedom is one less than four times the total number of days involved.

Sometime between day 11 and day 17 a fuse blew in the B pulser. Without any means of pinpointing the malfunction, the entire seven days' data for Bay B had to be eliminated from the study. The method of analysis described considered total plant catch as opposed to individual bay catches. Consequently all the data had to be reduced to the least common denominator, and thus the analysis was performed on the data for a 20 day period. Using all catfish species and eels as the response variable there are 79 degrees of freedom. These 70 degrees of freedom may be subdivided according to the design of the experiment as given in Table 4.

The data for all bays using catfish species and eels greater than 10 cm. in total length are shown in Table 5.

The pertinent question in this entire series of experiments is the significance of the variability due to the treatments, which are energized (X) and unenergized (Y) electrode arrays. Due to the double change-over nature of the experimental design the effect due to treatments is identical with the interaction between the cubic term and group (P_3 X G) of Table 4. That this is true can be demonstrated by the following 2 X 2 table.

Period cubic term

Group	1 + (3) III	(3) II + IV
A	$X_1 + 3\ X_3$	$3\ Y_2 + Y_4$
B	$Y_1 + 3\ Y_3$	$3\ X_2 + X_4$

The interaction between cubic term and group is determined from the diagonals of the above table. That is, the interaction is a function of the expression $(X_1 + 3\ X_2 + 3\ X_3 + X_4) - (Y_1 + 3\ Y_2 + 3\ Y_3 + Y_4)$. This expression is also the number of catfish and eels caught during the time the electrode array was electrically energized minus the number of the same species captured during the same period of time when the electrodes were unenergized, which is precisely the difference the experiment was designed to test.

Table 4. Effects of energized electrodes on bottom fish at CY, Bay A data. (Subdivisions of degrees of freedom specifying structure of experiment.)

	Group	Degrees of freedom	
Periods:			
Linear term	P_1	1	
Quadric term	P_2	1	
Cubic term	P_3	1	3
Days:			
Days in group A	D_A	9	
Days in group B	D_B	9	
Groups	G	1	19
Interactions:			
Between linear term and days in Group A	$P_1 \times D_A$	9	
Between linear term and days in group B	$P_1 \times D_B$	9	
Between linear term and group	$P_1 \times G$	1	19
Interactions:			
Between quadratic term and days in group A	$P_2 \times D_A$	9	
Between quadratic term and days in group B	$P_2 \times D_B$	9	
Between quadratic term and group	$P_2 \times G$	1	19
Interactions:			
Between cubic term and days in group A	$P_3 \times D_A$	9	
Between cubic term and days in group B	$P_3 \times D_B$	9	
Between cubic term and group	$P_3 \times G$	1	19
TOTAL			79

Table 5. Response to energized electrode (X) versus unenergized electrode (Y) showing numbers of bottom fish > 10 cm. and X + .05 transform.

Period I	Period II	Period III	Period IV	Differences
		GROUP A		
X_1 (ON)	Y_2 (OFF)	X_3 (ON)	Y_4 (OFF)	$-X_1 + 3Y_2$ $-3X_3 + Y_4$
5 (2.345)	11 (3.391)	12 (3.536)	13 (3.674)	0.894
13 (3.674)	17 (4.183)	7 (2.739)	19 (4.416)	2.186
6 (2.345)	14 (3.808)	5 (2.345)	12 (3.536)	5.375
5 (2.345)	10 (3.240)	3.(1.871)	7 (2.739)	4.501
2 (1.581)	3 (1.871)	2 (1.581)	2 (1.581)	0.870
3 (1.871)	2 (1.581)	1 (1.225)	3 (1.871)	1.068
3 (1.871)	3 (1.871)	6 (2.550)	2 (1.581)	-2.327
2 (1.581)	2 (1.581)	3 (1.871)	3 (1.871)	-0.580
1 (1.225)	1 (1.225)	2 (1.581)	7 (2.739)	0.446
3 (1.871)	4 (2.121)	5 (2.345)	2 (1.581)	-0.962
SUMS				
18.569	24.872	21.644	25.589	11.471
		GROUP B		
Y_1 (OFF)	X_2 (ON)	Y_3 (OFF)	X_4 (ON)	$-Y_1 + 3X_2$ $-3Y_3 + Y_4$
11 (3.391)	10 (3.240)	12 (3.536)	12 (3.536)	-0.743
6 (2.550)	9 (3.082)	15 (3.937)	7 (2.739)	-0.667
6 (2.440)	5 (2.345)	3 (1.871)	5 (2.345)	1.217
2 (1.581)	1 (1.225)	5 (2.345)	2 (1.581)	-3.360
3 (1.871)	0 (0.707)	3 (1.871)	5 (2.345)	-3.018
1 (1.225)	1 (1.225)	4 (2.121)	4 (2.121)	-1.792
1 (1.225)	3 (1.871)	6 (2.550)	2 (1.581)	-1.681
3 (1.871)	1 (1.225)	2 (1.581)	3 (1.871)	-1.068
2 (1.581)	2 (1.581)	2 (1.581)	6 (2.550)	0.969
3 (1.871)	2 (1.581)	2 (1.581)	0 (0.707)	-1.164
SUMS				
19.716	18.082	22.974	21.376	-11.306
BOTH GROUPS				
38.285	42.954	44.618	49.965	0.165

The error variance appropriate for testing the interaction between cubic term and group is obtained by pooling the interaction term and days in Group A with that between cubic term and days of Group B. Thus only the three portions of the total sum of squares which correspond to these three subdividions of the total number of degrees of freedom need be calculated. Using the third differences in the last column of Table 5, the calculations are:

Interaction between cubic term and days of group A ($P_3 \times D_A$):

$$\frac{(0.894)^2 + (2.186)^2 + \ldots\ldots + (-0.962)^2 - (11.471)^2/10}{1^2 + 3^2 + 3^2 + 1^2} = 2.517$$

Interaction between cubic term and days of group B ($P_3 \times D_B$):

$$\frac{(-0.743)^2 + (-0.667)^2 + \ldots\ldots + (-1.164)^2 - (-11.306)^2/10}{1^2 + 3^2 + 3^2 + 1^2}$$

Interaction between cubic term and group ($P_3 \times G$):

$$\frac{(-11.306)^2 + (11.471)^2 - (0.165)^2/2}{10\ (1^2 + 3^2 + 3^2 + 1^2)} = 1.297$$

From these data the relevant portion of the analysis of variance is developed as illustrated in Table 6. A significant difference between energized versus unenergized electrodes at the $\alpha = 0.05$ level has been demonstrated.

A similar analysis of the data for each of the bays was done on an individual basis. For Bay A the reduction was significant at the $\alpha = 0.05$ probability level, as was the reduction in impingement for Bay C. Using the same response variable as in the other analysis, no significant differences were demonstrated for the B and D Bay reductions.

A survey of the summary of results for all species, exclusive of bottom forms, is inconclusive. First, it is evident from an examination of the effects of the energized pulsers that the effect was inconsistent and relatively smaller than the results obtained for the bottom forms. When the data for all species greater than 10 cm. total length, exclusive of the catfishes and eels were analyzed statistically, none of the results of comparisons between energized and unenergized electrodes were statistically significant. The comparisons were made in a manner identical to those described previously and will not be described in detail.

Table 6. Analysis of variance results for the effect of the energized electrode array on bottom fishes for Phase I.

	d.f.	Sum of Squares	Mean Square
P_3 X D_A	9	2.517	
P_3 X D_B	9	0.978	
(P_3 X D_A) + (P_3 + D_B)	18	3.495	0.1942
P_3 X G	1	1.297	1.297
Total	19	4.792	

F = 1.297/0.1942 = 6.679*

Tab F_{05} 1, 18 - 4.41

PHASE II

Examination of the list of species encountered during Phase II reflects little variation from that illustrated in Table 1. Minor differences did occur; however, this is presumably attributable to seasonal variation, and not a consequence of any major changes in overall species composition of the plant environment. In the hopes of gaining greater insight for future decision making, the new data were analyzed with respect to species as well as groups. Subsequently, all the species-size classes were divided into groups based on similar behavior, ecology, and phylogeny. All clupeids except shad were grouped. The groups were then subjected to analysis identical to that for individual species These groups were 1) bottom fish: catfish and eels; 2) sunfish and related species; 3) clupeids; 4) perch; and 5) total fish. In contrast to the 10 cm. total-length delineation used in Phase I, a 15 cm. separation point was used in Phase II. At this time no systematic analysis has been performed to determine the best natural division. It is recognized that when division is made for the purpose of analysis, two organisms several centimeters apart may belong to separate classes, yet exhibit similar responses.

As indicated, data were collected at six hour intervals and the fish collected from each bay were recorded separately. The duration of both "on" and "off" cycles was two days. The survey began with the electrodes in the unenergized state with the goal of early establishment of a baseline.

Using each of the species or groups as the response variable, the data were analyzed by size class (i.e., < 15 cm., > 15 cm.). Preliminary analysis indicated no reductions in total numbers of fish. In most instances a substantial influx of fish was recorded during the energized periods.

To test whether there was a significant difference in the relative proportions of the various groups by bay, period, state, and day, the data were subjected to multivariate analysis. A significant difference in one or more factors or interactions occurred in only the following cases: grouped large clupeids, large shad, all other large clupeids, small clupeids, grouped large bottom fish, large white catfish, grouped large perch, large white perch, and grouped large sunfish. No significant differences were observed for any other species or group; in most instances the data recorded for these remaining fish were not significantly different from zero. The abridged results of the analysis after eliminating the extraneous species are shown in Table 7. Although the grouped large clupeids and the grouped

Table 7. Results of analysis of variance for selected species of groups.[1]

Source of Variation[2]	Degrees of Freedom	Large Shad	All Other Large Clupeids	Grouped Small Clupeids	Grouped Large Bottom Fish (Catfish and Eels)	Large White Catfish	Large White Perch	Grouped Large Sunfish
		SPECIES OR GROUPS						
B	3		*		**	**	**	**
P	3		**			*	**	
B x P	9							
D	1	*						
B x D	3							
P x D	3							
B x P x D	9							
S	1		**	*	**	**	**	**
B x S	3		*			*		
P x S	3	*	**			*	**	
B x P x S	9							
D x S	1		**	*			**	
B x D x S	3							
P x D x S	3		*					
B x P x D x S	9							

[1]Shown are only those groups which had significant results. Omitted for simplicity are all other species or groups. Only those blocks indicated had significant F values.
[2]B - Bay (A,B,C,D); P - Period (1,2,3,4); Day - (1st or 2nd); State - (on - off).
* significant at $\alpha = .05$
**significant at $\alpha = .01$

large perch showed significant results, they were omitted from the table as they merely reflected the magnitude of their principal component which was already included in the table.

As previously indicated, subsequent to the June survey several modifications were made; i.e., the pulse rate was increased, and chains were suspended from the negative electrode. At the same time the design was modified to investigate on and off periods of one week duration. The survey was conducted for one month under these conditions with the power remaining at earlier levels. For purposes of analysis this regime produced only two replicates; however, the sacrifice of replicates was deemed necessary in the interest of tuning the system under environmental time constraints.

It should be noted there that it was observed during this part of the survey that the screens did not come clean each time washed, and that a residual "population" remained. Any fish which did not appear fresh (i.e., had been passively collected by the plant along with other debris) was not counted. It was assumed that the occurrence of these residuals in the wash basket was random, or at least unbiased with regard to the survey, and consequently they were ignored. Early in September large quantities of debris forced the plant to initiate a schedule of continuous washing. Impingement data continued to be recorded under the established six-hour regime, but continuous washing precluded counts by individual bays.

During the month of September, only clupeids, catfish, perch and sunfish were caught in sufficient magnitudes to give any indication of treatment effect. The clupeids were omitted due to their extreme variability, which can in part be explained by their schooling behavior. Although the perch also exhibit this behavior, the size of the schools varies greatly. If we assume the schools are randomly distributed over the river, the small schools of the white perch would be much more uniformly distributed. Any incident of clupeid impingement is likely to be large due to school size, and sporadic due to the high degree of dispersion. Although schooling behavior may largely break down after dark, it would still leave large numbers of fish concentrated in a given area. With minimal replicates it was felt that consideration of this high variability data would be of little value, if not dangerous.

With the exception of one anomalous week of high clupeid counts, fish smaller than 15 cm. were rare for the June part of the survey. During the month of September juveniles began to appear in the wash, permitting the first consideration of this category. As with the June portion of the survey, all groups of fish larger than 15 cm. showed significant increases during the energized phase at

$\alpha = .005$. The small sunfish were reduced 59 percent and the small perch were reduced 65 percent. The small catfish demonstrated no significant response to the electrode state.

The significant reduction of some fish less than 15 cm. in length raised an interesting question. Were these small fish in a different physiological and/or behavioral state than the larger fish, or were they responding to different levels of stimulus? With little way to evaluate the first alternative due to time constraints, we chose to investigate the question of stimulus levels. If the field strengths present produced the desired effect in small fish, then the levels may have been too high for the larger fish.

Measurements of the electric field strength were made following the installation of the chains on the cathodes. Sampling electrodes mounted on a fiberglass pole were connected to a portable oscilloscope, permitting direct measurement of voltage gradients. Peak pulse values of 1.0 to 1.1 Volts per cm. were observed adjacent to both the anodes and the cathodes, with 0.3 to .5 V/cm. common midway between the electrodes and under the suspended cathode. Over the bottom-mounted anode, values of 0.04 to 1.0 V/cm. were measured from the surface to the bottom, but these values dropped to 0.02 to .06 V/cm. about 3 m. outside the anodes.

Unlike other investigations (Texas Instruments, 1973) where the large fish captured were in very poor condition, these fish were in surprisingly good condition, indicating that they were probably being overstimulated.

Consequently, the decision was made to decrease electrode voltage by $\sqrt{2}$ factor with resultant 50 percent drop in power. The results at this decreased setting (70 percent of initial levels) were even more encouraging. All three groups (catfish, sunfish, and perch) were significantly reduced in the smaller size category. The small catfish were reduced 89 percent in the energized state, and the perch and sunfish were reduced 52.5 and 67 percent respectively. The adults of the catfish and perch were again significantly increased; however, the larger sunfish were reduced, although not significantly. In view of the fact that large sunfish are the smallest of the over 15 cm. category, this reduction indicated more appropriate levels. Both size categories of sunfish combined had a 64 percent reduction in the energized state, significant at the $\alpha = .005$ level.

By 1 November, population levels and activity levels of animals present had declined seriously, resulting in limited daily catches. With the optimism that the tuning was taking the appropriate course, the first of the "mini-experiments" was begun.

The input voltage was again decreased by $\sqrt{2}$, reducing it to 50 percent of that used at the onset of Phase II. By this time the catfish were the only group present in sufficient numbers to undergo comparisons. Analysis was performed on catfish greater than 15 cm. and on total fish of both size classes. At this point the large catfish were significantly reduced 74 percent (α = .005). However, all fish exclusive of catfish were increased 43 percent (α = .05). When combined there was a 39 percent reduction in all fish captured greater than 15 cm.

With the results of "mini-experiment" even more encouraging the decision was made to drop the voltage once again by $\sqrt{2}$. This setting now placed the input voltage at approximately 36 percent of the initial values. Again only catfish and "all others" could be studied. Catfish specimens greater than 15 cm. were significantly reduced by 75 percent (α = .005). All fish exclusive of catfish were reduced by 31 percent; however, this figure was not significant. When combined, all fish greater than 15 cm. were significantly reduced 54 percent (α = .005) with the electrodes energized.

During the first mini-experiment there was no significant difference in the numbers of any small fish caught based on electrode state. During the second mini-experiment, fish less than 15 cm. were not caught in sufficient numbers to permit analysis.

DISCUSSION AND CONCLUSIONS

PHASE I

A total of 24 fish species were recorded during the experimental period of August 18 to September 15, 1972. This compares favorably with the list of resident species reported in the Connecticut River Ecological Study (Merriman, 1971). It is recognized that the 0.375 inch square mesh of the travelling screens of the intakes and a 0.50 inch mesh in the collecting basket were selective for the larger fish. Furthermore, it is also recognized that the effectiveness of the energized electrode arrays was relatively low for small fish for reasons stated previously.

The division of the fish species captured into two reasonably similar ecological groups seemed justified. The 10 cm. total length minimum size for inclusion in the data for analysis was arbitrary. However, this seemed reasonable on the basis of a preliminary examination of the raw data.

The data of Table 2 showed consistent reductions on the order of about 20 percent when the pulser was energized using catfish and eels as the response variable. Statistical analyses of the raw data for the total of the four bays tended to confirm the summary data of Table 2. When analyzed individually the results were inconsistent. Bays A and C showed significant reductions, but reductions in Bays B and D were inconclusive. However, no satisfactory explanation for this variability has been found.

An examination of the combined data for all catfish and eels (Table 3) demonstrated that the relative proportions captured by the energized versus unenergized electrodes were not similar by period. Period I seemed anomalous with a higher proportion of fish captured with the pulser on than when it was off. No adequate explanation for this is available from the observations. Since catfish and eels are nocturnal it seems possible that fish repelled during the night might have sought the shade and darkness of the bays more actively as daylight approached. However, no direct evidence for this is at hand.

At the same time, any investigation of the short term differences (i.e., between periods) must take into consideration the residual population living between the outer reaches of the intakes and the travelling screens. Any survey of differences without sufficient replicates may be extremely dangerous.

An examination of the reduced data for all fish species exclusive of catfish and eels showed more variability with completely inconsistent results. A statistical analysis identical to those performed previously demonstrated no significant differences in the numbers of fish captured when the pulsers were on compared to the off condition.

The reasons for obtaining relatively good reductions in the number of catfish species and eels with the pulsers on and no significant reductions of other species are not entirely clear. However, some tentative explanations are offered. First, it is postulated that some silting (sedimentation) occurred in the vicinity of the base of the electrodes which were imbedded in a block of wood in contact with the bottom. Sediment deposit on the block and electrode bases could have concentrated the current from each pulser in the sediment instead of the water column which may have a higher resistivity. Thus bottom dwelling fish would have been subjected to more intense electrical fields than the free-swimming forms. Secondly, it has been reported (Bennett, 1971; Rommel and McCleave, 1972) that catfish and eels are among a number of fish which have receptors that are specialized for the detection of weak electric fields. From the above reports it is suggested that the catfish and eels were more sensitive than the other species to the electric fields which were produced and that the major electric fields may have been in the vicinity of the sediments where these species are commonly found.

PHASE II

The species collected during the Phase II survey (June, September, October and November, 1974) compare favorably with the earlier study. Again the mesh of the travelling screens and collecting basket were selective for the larger fish. Based on examination of the raw data for the Phase II survey, a 15 cm. separation used in earlier studies was resumed.

The summary of data for June in Table 6 indicates significant differences for several of the factors. The difference between bays was not consistent. No satisfactory explanation for either a difference between bays or the lack of consistency between groups has been found. However, the bays have been shown to be significantly different from the point of catchability, independent of the presence of an electric field. Many variable factors among the bays, including silting, currents, bottom topography, differential clogging, and temperature can have the effect of varying the bays from a fish's standpoint. When considering the significant differences between periods, it is normal to observe differences reflective of natural rhythms and to expect an upset of any cycle with the introduction of an external stimulus. With the exception of electrode state, the remaining significant interaction terms were not deemed instructional with regard to the problem of tuning. Even in the absence of confirming analysis, visual inspection of the data confirmed a strong attractive response to the energized electrodes. Although opposite from the desired effect, this was an indication of stimulus response. From the onset, the possibility of an attractive response could not be ignored based on the extensive use of electricity in fishing. The extremely fresh condition of the fish indicated the possibility that they were being stunned by the field. By September enough juveniles were present to permit a comparison between size classes with respect to electrode state. The significant reductions in theperch and sunfish less than 15 cm., coupled with significant increases in the larger size classes, clearly indicated a size/voltage gradient problem. The smaller fish were receiving a lower total voltage over their length which elicited an avoidance response, while the larger fish were being stunned and passively collected by the plant. With more optimistic results from each power cutback, this conclusion gained substance. At the time of this writing tuning experiments are still underway.

Without question these studies indicate an advance toward the minimization of fish impinged.

Whereas initially it was felt that elevated voltage levels might merely extend the coverage of the field, it now seems clear that this was an erroneous assumption. Field strengths must be kept below a level at which the fish are stunned. This level cannot be predicted without a priori knowledge of various physical parameters such as conductivity, electrode spacing, material, etc., nor without some a priori knowledge of the biology of the species involved.

An extremely interesting aspect of this study is the potential for selectivity which the system offers. Based on seasonality, natural abundance or ecological importance, the system can be adjusted to minimize certain groups. With the installation of conduit electrode arrays in all four bays in 1971, the impingement of fish greater than 15 cm. of five species was reduced 68 to 82 percent over the 1968-69 levels (Hyman and Pikanowski, 1974). When this array was energized in 1972, the impingement of bottom fish again dropped more than 20 percent. Finally, following the tuning of the Phase II system the numbers of fish over 15 cm. of all species encountered was reduced 54 percent. Based on these results there seems to be little doubt about the feasibility of barrier systems. However, at the same time it is also recognized that the optimum system has yet to be reached.

LITERATURE CITED

ANONYMOUS. 1965. An experiment in electrical fishing with an electric field used as an adjunct to an otter trawl net. Commercial Fisheries Review Vol. 27, No. 6: 1-11.

BENNETT, M. V. L. 1971. Electroreception. In W. S. Hoar and D. J. Randall, Fish Physiology. Vol. V: 493-568.

BURKEY, H. T. 1917. Electric Fish Stop. U.S. Patent No. 1.269,380.

BURROWS, R. E. 1957. Diversion of adult salmon by an electric field. U.S. Fish and Wildlife Service. Special Scientific Report-Fisheries No. 246: 1-11.

ELLIS, J. E. and C. C. HOOPES. 1972. A freshwater fish electromotivator (FFEM): Its characteristics and operation. NOAA Technical Report NMFS Special Scientific Report-Fisheries No. 660 (U.S. Department of Commerce).

ERKKILA, L. F., B. R. SMITH and A. L. McLAIN. 1956. Sea Lamprey Control on the Great Lakes: 1953&1954. U.S. Department of the Interior. Fish and Wildlife Service. Special Scientific Report-Fisheries No. 175.

FEDERER, W. T. 1955. Experimental design. The MacMillan Company. New York. 544 pp.

HYMAN, M. A. M. and R. A. PIKANOWSKI. 1974. Statistical analysis of fish counts at Connecticut Yankee Atomic Power Company. Unpublished report submitted to Northeast Utilities Service Company.

JOHNSON, J. L. and R. L. McNEELY. 1966. Applications of Modern Technology in Commercial Fisheries. Geo-Marine Technology. Vol. 2, No. 9: 26-30.

LETHLEAN, N. G. 1953. An investigation into the design and performance of electric fish screens and an electric fish counter. Trans. Royal Soc. Edinburg. Vol. LXII, Part II, No. 13: 479-526.

MERRIMAN, D. (ed.) 1965-1972. The Connecticut River Ecological Study, Semi-Annual progress reports Nos. 1-14 to the Connecticut Water Resources Commission (Incorporated into the new Department of Environmental Protection, October 1971). Mimeo, ca 675 pp.

MEYER-WAARDEN, P. F., E. HALSBAND and I. HALSBAND. 1960. Bibliographie Uber die Elecktro fischerei und chre Grundlagen. Arch. Fischereiwiss. xi(1): 1-104.

MOORE, H. B. 1958. Marine Ecology. John Wiley & Sons, Inc. New York. 493 pp.

NIKONOROV, Ivan. 1964. Electrical fishing with lights and pumps. World Fishing. Vol. 13, No. 6: 52-55.

PEASE, N. C. and W. R. SEIDEL. 1967. Development of the Electro-Shrimp trawl system. Commercial Fisheries Review. Vol. 29, No. 8/9: 58-63.

ROMMEL, S. A. JR. and J. D. McCLEAVE. 1972. Oceanic Electric Fields: Perception by American Eels. Science. 176: 1233-1255.

SAILA, S. B. and C. E. WILLIAMS. 1972. An electric trawl system for lobsters. Marine Technology Society Journal. Vol. 6, No. 5: 25-31.

SMITH, E. D. 1966. Electrical Shark-Barrier Research. Geo-Marine Technology. Vol. 2, No. 10: 10-14.

TEXAS INSTRUMENTS. April 1973. Hudson River Ecological Study: Area of Indian Point. 1st Annual Report, prepared for Consolidated Edison by Texas Instruments Service Group.

VIBERT, R. (ed.) 1967. Fishing With Electricity. London. Fishing News (Books) Ltd. 276 pp.

WOODHEAD, A. D. 1963. Fish Behavior--The Migration of Fish. World Fishing. Vol. 12, No. 3: 37-39; Vol. 12, No. 4: 36-38.

GAS-BUBBLE DISEASE:
A REVIEW IN RELATION TO MODERN ENERGY PRODUCTION

R. E. Wolke
Marine Pathology Laboratory
University of Rhode Island, Kingston, Rhode Island 02881
G. R. Bouck
Environmental Protection Agency
National Water Quality Laboratory, Corvallis, Oregon 97330
R. K. Stroud
Department of Veterinary Medicine
Oregon State University, Corvallis, Oregon 97330

ABSTRACT

The literature is reviewed and gas-bubble disease (GBD) is considered in light of recent findings. The process is defined and the environmental parameters necessary for its development examined. The role of power plants in its pathogenesis is also considered. Special emphasis is placed on the description of gross and histopathological lesions as well as newer methods for field diagnosis employing gross lesions and rapid techniques for the *in situ* measurement of dissolved gas total pressure.

INTRODUCTION

Gas-bubble disease (GBD) of fish is a pathological process due to physical causes usually characterized by gas emboli (*pompholyhemia*), exophthalamus, systemic emphysema, disorientation and death. The term is also used in relation to a similar disease of invertebrates but in common usage refers to a condition of gas supersaturation affecting the fish. The nearest corresponding process in higher vertebrates is known as decompression sickness, high altitude sickness or the "bends".

The disease has been recognized in fish since 1898, when Gorham reported the condition in salt-water aquaria. Signs included emphysema of the cornea and fin epidermis as well as loss of equilibrium. Gorham recognized a possible bacterial etiology but was unable to isolate a causitive organism. He further observed that fish allowed access to deeper water were free from the disease. Reasoning the disease was therefore due to some physical imbalance, experiments were conducted relating gas-bubble disease to decompression sickness of man.

A more extensive description of signs, lesions, species affected and pathogenises was made by the same investigator in 1899. A series of articles followed recognizing the physical origin of the disease and reporting various gases as responsible for bubble production. Marsh and Gorham (1905) reported the disease among marine aquarium fish at Woods Hole, Massachusetts, incriminating nitrogen as the primary, if not the only, gas producing emboli. Marsh (1910) again reported an excess of nitrogen in conjunction with mortality in trout placed in water from a driven well. Early studies suggested two other gases, oxygen and carbon dioxide as possible causes of the disease (Plehn, 1924; Mrsic, 1933; Woodbury, 1942). The roles of the various gases, while not as yet fully resolved, were more completely defined by the work of Englehorn (1943) and Shirahata (1966). In 1913, the disease was reproduced in the laboratory by raising water temperatures without subsequent loss of gas, thus bringing about saturation of the environment (Shelford and Allee, 1913). This observation was to have an important bearing on the understanding of gas-bubble disease as it relates to modern energy production.

Water has long been used as a source of energy. It has been exploited for its kinetic properties both as a liquid and as a gas In addition, its high specific heat makes it an excellent cooling agent (Langbein, 1973). These two qualities of water; its kinetic potential and its high specific heat have lead to its use in steam and electric power generation and as the primary plant coolant in the production of nuclear energy.

Hydroelectric and nuclear plants are located near large bodies of water and call freely upon this natural resource to aid in the production of energy. The water is dammed, often caused to drop great heights, and frequently heated many degrees above ambient. These conditions may lead to supersaturation of the water with atmospheric gases which in turn result in pathological processes among the indigenous fauna.

We must recognize that production of energy, while of great importance to man, is also a cause of disease and death among fresh and salt-water fish. There is little question that the utilization of energy has increased exponentially within the past 50 years and will undoubtedly continue to increase during the next 50. Nuclear plants will become more efficient as breeder reactors, but there may be increasing amounts of heat dissipated into water resources (Perrin, 1971). More dams will also be built. As energy needs grow and are fulfilled, so also will physical and thermal pollution grow resulting in a probable increase of fish kills due to gas-bubble disease. It is not only thermal pollution *per se* which will play a role in evaluating environmental impact of energy production, but also its secondary effect, that of gas supersaturation.

It is the purpose of this paper to review gas-bubble disease as it relates to fisheries and modern energy production and to consider newer methods of physical and anatomical pathological diagnosis. A recent review of the gas-bubble disease world literature was prepared by Weitkamp and Katz (1973).

PHYSICAL FACTORS

The solubility of atmospheric gases within a liquid are dependent upon the nature of the gases and the liquid, the pressure of the gases and the temperature. Gas content of the liquid may be expressed in ml/l; however, this does not take into consideration varying temperature and pressure. A more efficient method is to express the content as percent air saturation within moist air at a pressure of 760 mm. Hg (Rucker, 1972).

Solubility of gases within a liquid is increased as the surrounding pressure is increased. Conversely, when the surrounding temperature is increased the solubility of the gases within a liquid are decreased. Because of this pressure relationship, the saturation levels of nitrogen and oxygen, at a constant temperature, will increase in direct relation to the depth of the water. At sea level and 10°C saturation levels for nitrogen in water are 14.61 ml/l and for oxygen 7.89 ml/l. However, a saturation level of either of these gases in water at 10 meters depth and 10°C (N_2 = 27.0 ml/l; O_2 = 15.5 ml/l) would result at sea level in a supersaturation of 151 percent nitrogen and 150 percent oxygen.

Similarly, the solubility of nitrogen and oxygen in water at a relatively constant pressure is inversely proportional to water temperature. At 5°C and 1 atmosphere the saturation for nitrogen in distilled water is 16.26 ml/l while at 14°C it is 13.34 ml/l (Weiss, 1970). Therefore, the sudden heating of a body of water may result in a momentary supersaturation prior to equilibration.

The laws which govern the solubility of atmospheric gases in water apply as well to their solubility in blood and body tissues (Harris, *et al*., 1945; Harvey, *et al*., 1944a). The partial pressure of nitrogen in tissue (569 mm Hg), however, is far greater than that of oxygen (20 to 40 mm Hg) which accounts for the significant amounts of nitrogen involved in the production of gas emboli (Nims, 1951). It follows, therefore, that the gaseous saturation of the blood and tissue of fish will be in equilibrium with the surrounding water and rapid changes in external saturation levels may result in similar changes in vivo.

The role of individual gases (such as oxygen and nitrogen) is not understood fully at this time. In isobaric testing of total dissolved gas pressure, fish tolerated hyperbaric oxygen longer and perhaps better than hyperbaric nitrogen (Rucker).[1] In this regard, the decompression period of skin divers can be reduced 50 percent when half of the nitrogen is replaced with helium in the breathing air (Workman, 1963). These observations indicate that each gas with its unique properties can impact the situation to some extent. But regardless of the gas composition, gas-bubble disease cannot occur unless the total dissolved gas pressure (i.e., the sum of their individual tensions) exceeds all compression forces including the pressures of air, water, blood and tissue elasticity (Anon., 1972; Webster, 1955; Doudoroff, 1957).

ENVIRONMENTAL FACTORS

In order for the signs of gas-bubble disease to develop, the tissues and blood of the affected fish must be supersaturated with atmospheric gases, especially nitrogen. This in vivo supersaturation is related to the supersaturation of the surrounding environment and may result in a marked difference between the solubility of surrounding gases and those of the fishes interior milieu. These in vivo gases, if present as supersaturates will seek equilibrium with the atmosphere or with the surrounding water. It is in the process of seeking equilibrium that pompholyhemia and emphysema occur.

If one considers varying temperatures and pressures and relates these factors to movements of fish in the environment, it is possible to envision a number of theoretical interactions under which gas-bubble disease might arise. These interactions may be classified in two major groups. In the first, the fish is essentially static and the surrounding environment becomes supersaturated, while in the second, the fish is mobile and moves into a supersaturated environment.

---

[1]Rucker, R. R. Unpublished manuscript. Variation in oxygen-nitrogen ratio with total gas pressure. Bureau of Sport Fisheries and Wildlife Western Fish Disease Laboratory, Bldg. 204, Sand Point Naval Support Activity, Seattle, Washington 98115.

The static fish-changing environment interaction occurs with greatest frequency under aquaculture conditions where it is difficult or impossible for the fish to move away from the supersaturation. Reports are conflicting concerning fish avoidance of supersaturated areas. Shelford and Allee (1913) found no indication that fish are capable of detecting supersaturated areas. Ebel (1969) further reported that fish will not attempt to seek hydrostatic pressure compensation. However, Meldrim, et al.[2], in laboratory experiments reported species differences in avoidance of supersaturation. In their studies, the perch was relatively unresponsive to supersaturation while the golden shiner and satinfin shiner were very responsive at controlled temperatures. Avoidance reactions by the golden shiner, however, were overridden by temperature preference when temperatures were increased 5 to 10°C (Meldrim, et al.). In experiments conducted to determine the effect of temperature increase on fish already in supersaturated water, mortality rates were decreased among fish kept in a deep tank and allowed to move to depths as great as 9 meters (Ebel, et al., 1971). Dawley, et al.[3] in long-term supersaturation exposure experiments with chinook salmon found that the average depths of fish tested in deep tanks increased with increased saturation. Similarly, May[4] reported 100 percent mortality in whitefish and trout held in cages just below the surface of supersaturated water but the same fish held in cages allowing access to 10 feet had a maximum of 70 percent mortality.

---

[2]Meldrim, J. W., J. J. Gift and B. R. Petrosky. Unpublished manuscripts. Responses of several freshwater fishes to water containing various levels of supersaturation. I. A comparison of two different methods for achieving supersaturation. II. The effect of increased temperature associated with gas supersaturation on the response. Ichthyological Assoc., 100 S. Cass Street, Middletown, Delaware 19709.

[3]Dawley, E. M., M. Schiewe and T. Newcomb. Unpublished manuscript. Effect of long-term exposure to supersaturation of dissolved atmospheric gases on juvenile chinook salmon and steelhead trout in deep and shallow test tanks. National Marine Fisheries Service, Northwest Fisheries Center, 2725 Montlake Blvd. E., Seattle, Washington 98102.

[4]May, B. Unpublished manuscript. A preliminary evaluation of the effects of gas-bubble disease on fish populations in Kootenai River below Libby Dam. Montana Department of Fish and Game, Rt. 1, Box 1270, Libby, Montana 59923

The mobile fish-supersaturated environment interactions arise with greatest frequency in the natural environment in relation to natural phenomena or human interference interacting with natural movements of large numbers of fish. The natural phenomena include such variables as sunlight, water turbulence and waterfalls while human interference includes thermal effluents and dammed falling water.

There are basically three mechanisms by which water may become supersaturated. These mechanisms are air entrapment, change in temperature, or change in pressure.

Air entrapment.--Air entrapment or entrainment, while closely related to changes in pressure, is a process in which gases are distributed in excessive amounts throughout a liquid because the gas is surrounded by the liquid and forced into solution subsequent to a sudden rise in pressure. A classic and relatively common example of the process is the damaged water pump hose which at each pump pulsation, sucks air into the passing water. In 1904, Marsh and Gorham reported mortality in marine aquaria at Woods Hole, Massachusetts, in conjunction with a porous wooden suction pipe. They attributed supersaturation to entrapped air subjected to increased hydrostatic pressure just beyond the pump. More recent investigations have confirmed entrapment as a source of supersaturation within aquaria (Dannevig and Dannevig, 1950; Harvey and Smith, 1961; Spotte, 1973). Similarly, intake pipes, the mouth of which are not totally below the surface may be responsible for air entrapment (Westgard, 1964).

Supersaturation of rivers in the Pacific Northwest is well documented (Ebel, 1969 and 1971). Air entrapment is an important source of this supersaturation and has been incriminated as a cause of gas-bubble disease in migrating salmon. The most frequent cause of air entrapment is cascading water from hydroelectric dams and spillways which drive the captured air to a depth sufficient to result in a solubility within the water greater than their solubility at sea-level. Both Harvey and Cooper (1962) and Westgard (1964) have reported gas-bubble disease in supersaturated river water secondary to air entrapment in plunge basins. Merril, et al. (1971) reported salmon mortality in association with high spillway flows near the Bonneville Dam, Columbia River. Spills at the John Day Dam, Columbia River have resulted in dissolved nitrogen levels of from 123 to 143 percent supersaturation (Beiningen and Ebel, 1970).

Smith (1972) has shown that this supersaturated water tends to remain supersaturated because it runs in the deeper regions of the river and is recharged at subsequent spillways. Supersaturation levels as high as 116 percent have been measured at Astoria, Oregon, over one hundred miles downstream from Bonneville Dam.

Air entrapment may also occur within turbines operating at reduced levels. MacDonald and Hyatt (1973) report two fish kills below a power station in association with nitrogen levels as high as 125 percent. Power personnel had allowed air to enter the turbine system to reduce negative pressures associated with lowered water flow.

Temperature.--A second mechanism resulting in supersaturation is temperature increase. These temperature increases are often sudden and of both natural and human origin. Within aquaria (static fish--changing environment) there is experimental evidence that rapid increases in temperature will result in supersaturation and gas-bubble disease (Shelford and Allee, 1913; Engelhorn, 1943; Malouf, *et al.*, 1972).

Increases in temperature as a cause of supersaturation also play a role in the natural environment. Under these circumstances, however, the production of gas-bubble disease is dependent upon susceptible fish migrating into the supersaturated area or remaining in the area. The mechanism of supersaturation via temperature increase has the added liability that many species of fish are attracted to areas of higher temperature. The temperature increase, therefore, may not only be responsible for a lethal environment, but also for luring the unsuspecting victim to that environment.

In 1963, Renfro reported a natural mortality of fish in Galveston Bay, Texas, associated with climatic conditions resulting in increasing water temperatures and intense photosynthesis. While nitrogen values were not determined, oxygen supersaturation was 250 percent. Harvey (1964) reported gas-bubble disease in sockeye salmon in conjunction with a warming of a lake surface. He noted that nitrogen saturation was higher than oxygen below the thermocline. Calculations indicated that during the month of June fish migrating to the surface from any depth below nine meters would have approached a supersaturation of 130 percent. This same investigator reported in 1967 that lake water in the region of the thermocline may become markedly supersaturated as a consequence of warming during the spring and summer. Often the supersaturated water is not subjected to surface mixing and gas may be held in solution by hydrostatic pressure.

In a discussion of supersaturation of the Gulf of Maine coastal waters, Stickney (1968) suggested that protected, shallow embayments on bright sunlit days might give rise to rapid water warming with subsequent supersaturation. The authors also discussed the possibility of supersaturation arising as a consequence of tides which mix large quantities of warm and cold water. While this situation is certainly theoretically possible, it is probably of little consequence.

Temperature increases of human origin are related to the cooling of steamelectric and nuclear power plants. Thermal effluents from these plants may be far above ambient temperatures. Supersaturation in relation to heated discharges is a real problem (Nakatani, 1969). Other investigators have also recognized this problem noting differences as great as 15°C between effluent and ambient temperatures (Ebel, 1969; Becker, 1970). Demont and Miller (1971) were the first investigators to report gas-bubble disease in the heated effluent of a steam generating station. Unfortunately, saturation levels for nitrogen and oxygen and the temperature differences causing the supersaturation were not reported. Further evidence for the incidence of gas-bubble disease in thermal effluents was supplied by Marcello and Strawn (1972). These investigators attempted to culture some marine fishes in the discharge canals of a steam-electric generating station in Galveston Bay, Texas. While species differences were noted, croaker, pompano and pinfish had clinical signs of gas-bubble disease including exophthalmus and abdominal rupture. These signs developed at a ΔT of 7 to 10°C.

In April of 1973, a mass mortality of menhaden was observed in relation to the thermal effluent of the Pilgrim Nuclear Power Station, Plymouth, Massachusetts (Marcello and Fairbanks).[5] One of the authors (REW) conducted necropsy examinations on a number of recently dead and moribund fish. Gross and microscopic lesions were compatible with a diagnosis of gas-bubble disease.

Effluent temperatures were as much as 15°C (27°F) above ambient. It is of interest to note that menhaden are a strongly schooling fish which prefer warmer waters.

Pressure.--A third mechanism responsible for supersaturation is change in pressure. Theoretically, this change might occur either to the environment (static fish--changing environment) or to the fish (mobile fish--supersaturated environment). In the first instance, water under pressure from deep wells is pumped to the

---

[5]Marcello, R. A. Jr. and R. B. Fairbanks. Unpublished manuscript. Gas-bubble disease mortality of Atlantic Menhaden, Brevoortia tyrannus, at a coastal power plant. Boston Edison Company, Nuclear Engineering Department, 800 Boylston Street, Boston, Massachusetts 02199.

surface resulting in a sudden decrease in pressure and supersaturation. Marsh and Gorham (1905) reported mortality of trout placed in supersaturated water from a spring feeding the Fisheries Station at Erwin, Tennessee, and a similar problem from artesian wells in Nashua, New Hampshire. Again, in 1910, Marsh reported water from a well driven in 1906 which was deficient in oxygen and contained excessive nitrogen and carbon dioxide. Trout placed in water from the well suffered acute mortality. Further reports of supersaturated spring water include those of Rucker and Hodgeboom (1953) and Matsue, et al. (1953).

RELATIONSHIP TO POWER GENERATION

As has been discussed, there are several ways that power generating stations can cause water to become supersaturated with atmospheric gases. Regardless of the cause, the result is water containing more dissolved air pressure than the atmosphere over the water. This causes gas bubbles to form on or within aquatic organisms, hence causes gas-bubble disease which can injure or kill fish. This problem already exists across the nation and around the world, both as a natural phenomenon and as a result of man's activities. Fortunately, it can be prevented at power generating stations providing there is knowledge and awareness of the potential problem. Therefore our purpose here is to review the cause and effect relationahip between gas-bubble disease, supersaturation, and power production, so that one can avoid or correct this water quality problem.

Gas-bubble disease has probebly existed since time immemorial. Scientific observations on gas-bubble disease date back at least 100 years to Bert's (1873) report and it is safe to say that the scientific community has learned much about this problem. On the other hand, little of this knowledge has been collected under natural circumstances that are directly applicable to power generating stations. Furthermore, gas-bubble disease has a somewhat unpredictable impact, probably because it is the product of so many biological and physical variables. Hence while we know enough to avoid this problem, there is still much that could be learned and might impact favorably the operation of power generating stations.

Supersaturation or hyperbaric dissolved total gas pressure might result at a power generating station for two primary reasons: (1) because air and water are in contact at hyperbaric conditions, and (2) because water is being warmed without adequate aeration. Examples of the former include air injection into turbines, entrainment of air at spillways or sluceways, venturi action of pumps and pipe joints or fixtures, and possibly deep water reaeration.

In each of these examples, the air and water are combined under pressure long enough to dissolve the air addition, following which the water leaves the system and is supersaturated relative to air pressure.

Heating of water causes supersaturation by increasing the kinetic energy of dissolved gas molecules, which in turn increases the total pressure that they exert. Examples of this include solar heating of impoundments or streams, and the rejection of waste heat into surface waters. Such heat induced supersaturation can present a serious problem, but usually on an intermittent basis. This is because the supersaturated effluent may be more preferable in temperature than the surrounding water, and therefore attract fish during colder periods. Fish kills of this type have been reported by DeMont and Miller (1971).

Once the water has been supersaturated with air, the result is a condition that is much more stable than one might ever expect to occur. For example, Columbia River spillage at Bonneville Dam has a saturation of about 1.25 atm. and arrives 200 km. down stream at about 1.18 atm. (Beiningen and Ebel, 1971). This is because it takes energy and time to cause supersaturation and desaturation. Furthermore, as the level of supersaturation diminishes, progressively higher levels of energy or time are required to reach the equilibrium state.

Control over dissolved gas composition seems to offer little practical value at the typical power generating station mainly because the magnitude of the problem is overwhelming. Such treatment costs could be very large, but its benefits would remain of questionable long-range value and it would fail to eliminate the root source of the problem. However, knowledge of the dissolved gas composition does have benefit indirectly because it helps to incicate the severity and possible cause(s) of the problem.

SIGNS AND LESIONS

The signs and general lesions of gas-bubble disease (GBD) have been described under field and experimental conditions by various authors (Gorham, 1901; Marsh and Gorham, 1905; Woodbury, 1941; Rucker and Hodgeboom, 1953; Egusa, 1959; Wood, 1968; Renfro, 1963; Shirahata, 1966; Beiningen and Ebel, 1970; Bouck, *et al*.[7]; Wyatt and Beiningen, 1971; Elling and Ebel, 1973).

---

[7]Bouck, G. R., G. A. Chapman, P. W. Schneider, Jr. and D. G. Stevens. Unpublished manuscript. Observations on gas bubble

The signs and lesions associated with GBD are dependent on many factors. Primary factors are the level of supersaturation and the duration of exposure, which in turn can be influenced further by water temperature, muscular activity, depth of water and others. The species of fish and its life stage are also important in the development and incidence of some types of lesions. In fish exposed continuously to sublethal levels of supersaturated water for long periods of time, the presence of bacterial and fungal secondary invaders and ectoparasites may complicate the otherwise typical picture of GBD.

Signs.--The signs observed in fish exposed to supersaturated water have been noted as loss of equilibrium, abnormal buoyancy, clinical blindness, bucal reflex or back flushing of gills in response to gill irritation, lethargy, narcotization and terminal convulsions (Gorham, 1901; Shirahata, 1966; Wyatt and Beiningen, 1971; personal observation, RKS). Fish frequently die with gills flared, a common sign of anoxia. Yolk sac fry will float upside down on the water surface due to gas accumulation within the yolk sac (Rucker and Hodgeboom, 1953; Wood, 1968). Bubbles of gas within the gastrointestinal tract of herring larvae and larval flat fish has also been reported to cause abnormal bouyancy (Dannevig and Dannevig, 1950; Henly, 1952). A sudden release of gas from the swim bladder of physostomus fish may be observed. Abnormal distention of the gas bladder may occur in both physostomus and physocliestus fish (Egusa, 1959; Shirahata, 1966; personal observation, RKS). The response to external stimuli by the lateral line system is greatly diminished or may completely disappear in fish under supersaturated conditions (Schiewe and Weber).[8]

Lesions.--Although GBD has been reported from many species of fish, the pathological effects of exposure to supersaturated water has been most extensively studied in salmonid species under laboratory or hatchery conditions which are inherently constant exposure shallow water situations as opposed to a more intermittent type of exposure experienced under natural conditions. Lesions may

---

disease in adult Columbia River sockeye salmon (Oncorhynchus nerka). Pacific Northwest Lab., Federal Water Quality Administration, Corvallis, Oregon 97330.

[8]Schiewe, M. H. and D. D. Weber. Unpublished Manuscript. Effect of gas bubble disease on lateral line function in juvenile steelhead trout. National Marine Fisheries Service, Northwest Fisheries Center, 2725 Montlake Blvd. E., Seattle, Washington 98102.

be directly or indirectly traced to two basic pathological processes. The occurrence of gas emboli within the blood vascular system is the primary pathological process. These emboli impede the normal flow of blood by blockage of capillary beds. The occurrence of "notable quantities of free gas" in the blood vessels and heart was described by Marsh and Gorham (1904). Other investigators have since made observations of gas emboli in intercostal, mesenteric, coronary, meningeal, testicular and ovarian vessels. The afferent gill vessels and the atrium and conus arteriosis are the most obviously affected (Woodbury, 1941; Renfro, 1963; Bouck, et al.[7]; Wyatt and Beiningen, 1971; personal observation, RKS). If the liver, kidney, or spleen is excised and squeezed, blood containing significant amounts of air bubbles may be obtained (Bouck, et al.[7])

Gill edema, degeneration, necrosis, aneurysms, and hemorrhage has been described as being associated with the blockage of afferent gill vessels with gas emboli (Pauley and Nakatani, 1967; Woodbury, 1941; Wyatt and Beiningen, 1971). Invasion of the necrotic gill tissue by the fungus Saprolengnia sp. has also been observed (unpublished observations, RKS and GRB).

Congestion of the mesenteric vessels, spleen, kidney and liver result from the plugging of the heart, ventral aorta and afferent gill vessels with emboli in adult chinook salmon (Figs. 1 and 2). Petechial hemorrhage throughout the body has been observed in fish surviving long-term exposures to supersaturated water. Although this hemorrhage may be due to capillary breakdown caused by prolonged congestion or direct vascular damage from gas emboli, it may also be due to systemic bacterial infections that may occur under the conditions of extreme stress caused by supersaturated water. Mesenteric and kidney edema and ascites observed in adult chinook salmon after prolonged exposure to sublethal saturation levels may have a similar pathogenesis. The physiological process of emboli formation in fish is still undetermined. Theories have been advanced based on mammalian research, but this aspect of the pathogenesis of GBD in fish has not been thoroughly investigated.

The formation of gas bubbles (cavitation) within a liquid or tissue appears to be a complex subject. Hemmingsen (1970) was unable to produce cavitation of water at atmospheric pressure with supersaturations of 140 atmospheres oxygen and 170 atmospheres nitrogen. Hills (1967) has shown that bubbles will form at the interface of two immiscible liquids such as lipid-water, tissue-water or glass-water boundaries. Earlier work by Harvey, et al. (1944a; 1944b) and Harvey (1951) investigated bubble formation in animals. These investigators concluded that bubbles form in

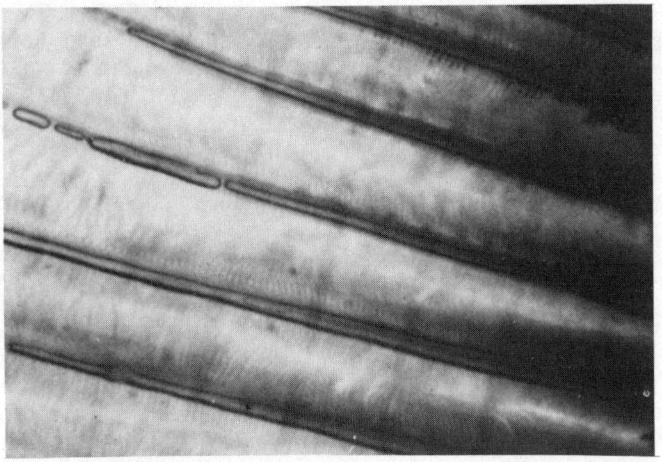

Figure 1. Gas emboli accumulation in the afferent lamellar vessels. X 10.

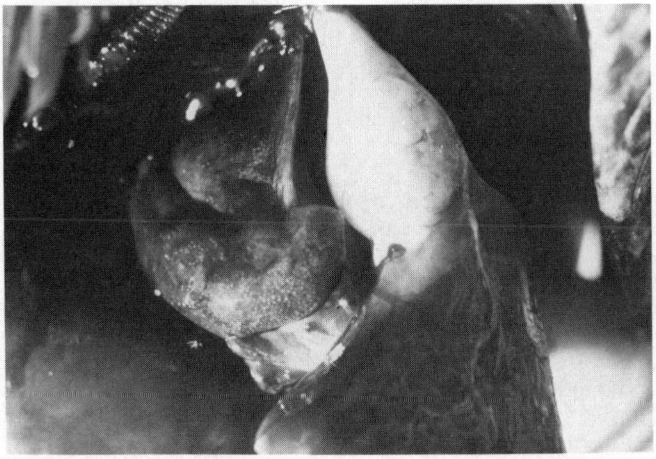

Figure 2. Atrial distension with mixture of gas emboli and blood in adult chinook salmon.

pre-pressurized connective tissue as a result of high local negative pressures from mechanical tension. Bubbles tend to form in tissue when it is pulled apart or excised. Evans and Walder (1969) investigating bubble formation in shrimp report that gas micronuclei must exist for cavitation to occur. Such gas nuclei would arise if gas solubility or surface tension were decreased.

Since exercise would increase tissue mechanical tensions and hence negative pressures it would follow that exercise might promote bubble formation. Whitaker et al. (1945), showed this to be the case in both frogs and rats. Bubbles appeared in veins leading from active muscles in animals decompressed to simulated altitudes while exercising. Similar experiments were conducted by Harris et al. (1945).

The relationship of bubble formation to fat is also of interest. Nitrogen is more soluble in fat than other atmospheric gases and decompression sickness appears to have a higher incidence among fatter individuals (Gersh and Catchpole, 1951). Bubbles are also more common in fat, adrenal gland and myelin sheaths than in muscle, liver or tendon of higher vertebrates (Gersh, et al., 1944). The relationship of these findings to the incidence of gas bubble disease in "fat" versus "lean" fish has not been investigated.

The second basic pathological process is the abnormal accumulation of gas within the body tissue exclusive of the blood vascular system (emphysema). Emphysema may be the result of gas cavitation or movement into tissues either from the blood vascular system, passage of gas across epithelial membranes from the external environment, or both.

Emphysema is most consistently seen in the fins and tail. The appearance of "gas blisters" or emphysema between the fin rays has been widely reported by many investigators as "external symptoms." These may appear singly as a series of small vesicles or may, on occasion, coalesce into columns extending the entire length of the fin ray. The size varies from 0.25 mm. diameter to vesicles 5 mm. or greater in diameter in larger fish. Individual fish may develop emphysema within any or all of the fins and tail. It should be noted that fish may die before emphysema develops, but if emphysema is noted, it may be considered as pathognomonic for GBD. The incidence of external symptoms, namely emphysema of the fins and body in fish exposed to Columbia River water with saturation levels from 115 percent to 125 percent, was compared to mortality by Blahm, et al. (1973). Eighty percent of juvenile chinook salmon dead of GBD held in 2.5 meter-depth tanks developed emphysema of the fins and body, while only 50 percent of juvenile coho salmon held in similar cages showed the same lesions. Seventy-nine percent and 96 percent incidence of lesions were seen, respectively, in fish held at 1 meter depth. At least 10 percent of 129 adult sockeye salmon trapped in the fish ladder at Bonneville Dam in 1969 had externally evident emphysema. In 1972, 47 percent of 55 adult sockeye, 40 percent of 5 adult steelhead and 26 percent of 35 American shad trapped had externally evident emphysema while no emphysema was seen on lamprey eels caught at the same time. A study on the incidence related to location in the above adult sockeye revealed that emphysema was located as follows: caudal fin, 72 percent; dorsal fin, 50 percent; pelvic fins, 44 percent; ventral fin, 20 percent; anal fin, 8 percent; and pectoral fin, 4 percent (Bouck, et al.[7]). A similar study on the incidence of fin emphysema in laboratory-exposed steelhead smolts was inconclusive. Gas vesicles in the tail of adult chinook salmon were less likely to form than vesicles on the fins (unpublished data, RKS).

Histologically, the interray tissue is observed to be greatly distended with gas (Fig. 3). The gas accumulation is not at the dermal epidermal junction. Lymphatics or venules appear to fill with gas and become grossly distended compressing the interrary connective tissue outward against the epithelium. Fibrin and some white blood cells may be observed to accumulate within the gas blisters. Rupture of the gas blisters may occur with secondary invasion of the lesion by Saprolegnia sp. and other fungal and bacterial species. Physical distortion of the cartilgenous fin rays may also be observed. Hyperemia and occasional hemorrhage at the base of pectoral, pelvic and anal fins may be observed in fish held in sublethal levels of supersaturated water. Fin and tail emphysema regresses within a few days if fish are removed from supersaturated water (Blahm, et al., 1973). According to Coutant and Genoway (1968), external gas blisters disappeared from fish 24 hours after death. Adult chinook salmon exposed to 120 percent supersaturated water retained emphasematous lesions in the fins and tail for 35 hours after death if they remained in the supersaturated water (unpublished observations, RKS). Persistance of emphysema post mortem is probably a function of tissue elasticity and the duration of the lesion.

Emphysema is also commonly reported along the external and internal sides of the posterior edge of the operculum (Fig. 4). The branchiostegal rays may be affected also. Opercular emphysema does not develop as rapidly and frequently in steelhead smolts as does the emphysema of the fins. In adult chinook salmon held in 115 percent supersaturated water for long periods, opercular emphysema was frequently accompanied by tissue edema and petechial hemorrhage.

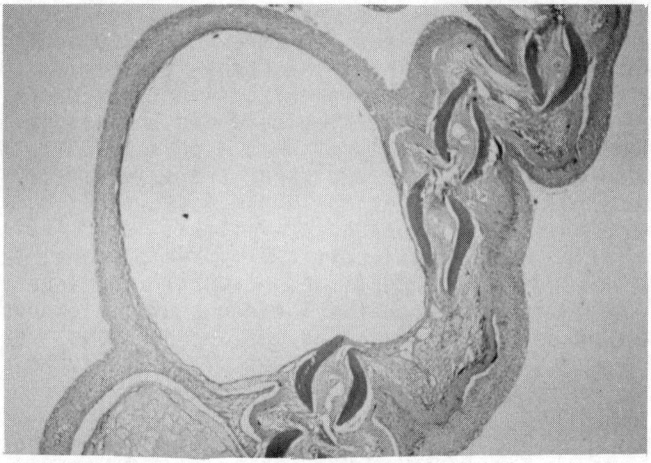

Figure 3. Cross-section of fin with extensive accumulation of gas in interray tissue. Note fibrin filled gas vesicle toward bottom. H. & E. X 45.

Figure 4. Emphysema of the orbit, operculum and skin of the head in steelhead smolt exposed to 115 percent supersaturation levels for a prolonged period.

Subcutaneous emphysema has been reported from the skin of the head, the oral cavity and gill arches, and along the body, especially in the region of the lateral line (Fig. 4). The roof of the mouth may contain gas-filled vesicles large enough to interfere with normal feeding and respiration. Hydropic degeneration of the epithelium and edema and hypertrophy of the connective tissue were described as pathological changes in the roof of the mouth that are unique to GBD and could be used as a diagnostic characteristic for detecting GBD in its early stages (Pauley and Nakatani, 1967). However, steelhead trout smolts exposed to 110, 115, 120 percent levels of supersaturation for up to 14 days did not have this lesion when examined histologically (unpublished data, RKS).

The occurrence of small bubbles of gas within the mucus layers of the skin has been reported (Marsh and Gorham, 1904). The net effect of these gas vesicles was to tear the protective mucus coating from the fish. Fish have also been observed as encased in an envelope of bubbles adhering to the external surface (Shirahata, 1966). These external bubbles would fill the oral cavity and opercular slits of fry, causing suffication.

In adult salmonids, small gas bubbles 0.5 to 2 mm. in diameter could occasionally be seen to lie directly over the surface of the scale under the epidermis (personal observations, RKS). Scales raised by subcutaneous emphysema may give the fish a roughened appearance. Steelhead trout subjected to 118 percent supersaturated water for 2 to 4 hours developed emphysema along the lateral line evidenced by a slight lifting of scales due to gas accumulation in the scale pocket. Normal function of the lateral line system was disrupted (Schiewe and Weber[8]). Areas of the body with extensive subcutaneous emphysema were prone to invasion of the devitalized tissue by fungus after prolonged exposure to sublethal levels of supersaturation.

"Popeye" or exopthalmia has been reported frequently in fish exposed to supersaturated water. It should be stressed, however, that exopthalmia is a lesion that occurs with many conditions other than exposure to supersaturated water. However, in GBD, close examination may reveal that the retrobulbular fatty tissue has become emphysematous forcing the globe of the eye from the socket. The palpebral conjuctiva may also be forced outward as the retrobulbular space becomes distended with gas (Fig. 4).

Exopthalmia was present in juvenile salmonids, but was not observed in adult fish exposed to the same levels of supersaturation (personal observations, RKS). Exopthalmia in itself does not seem to cause loss of sight and appears reversible if fish are removed from supersaturation conditions. However, protrusion of the eye often invites attacks from fish within the group, In spite of the attention given this lesion in the literature, the incidence of exopthalmia in juveniles is relatively low compared to the occurrence of fin and body emphysema. No damage to the optic nerve or the functional tissues of the eye has been reported in the literature. Exopthalmia is not a common lesion in adult salmonids. Hemorrhage into the anterior chamber of the eye, corneal erosion and finally rupture of the lens has been reported in adult chinook and sockeye salmon passing Columbia River dams during periods of high supersaturation (Beiningen and Ebel, 1970; Bouck, et al.[7]). Experimental exposure of adult chinook salmon to 120 and 115 percent supersaturated water conditions for as little as 70 hours and 238 hours, respectively, produced an accumulation of gas bubbles within the choroid plexes (Fig. 5). Hemorrhage, retinal detachment and release of free gas into the chambers of the eye resulted from this post retinal gas accumulation. In addition, the lens of the eye was observed in some fish to be pushed against the cornea and finally rupturing the lens from the eye through the eroded cornea (personal observations, RKS). Similar eye lesions in adult chinook and sockeye salmon passing Bonneville Dam have been reported by observers in fish counting stations (Beiningen and Ebel, 1970). The loss of lenses and complete eyes in adult chinook and sockeye salmon was reported in fish held in supersaturated river water for up to 44 days (Bouck, et al.[7]).

A unique lesion observed only after prolonged (15 days) exposure of adult chinook salmon to 115 percent and 120 percent supersaturated water was muscle cavitation (Fig. 6). Cavitation in the muscle undoubtedly occurs sooner but is unobservable until the muscle tissue has been stretched for a sufficient period of time to retain the cavity when the tissue is cut.

FIELD DIAGNOSIS OF GBD

The identification of gas-bubble disease is dependent upon evidence that the organism actually was in supersaturated water, developed emboli, emphysema or collaterla lesions. According to the text book, it should be relatively easy for anyone to identify gas-bubble disease, but of course this is not the case. Most published articles do not deal with the intermittent exposure that wild fish typically receive. Rather, the literature emphasizes lesions and signs that generally occur only after prolonged exposure at tolerated but adverse levels. The result is an emphasis upon externally evident signs such as exopthalamous (popeye) and emphysema (bubbles) in and under the skin, especially in transparent fins.

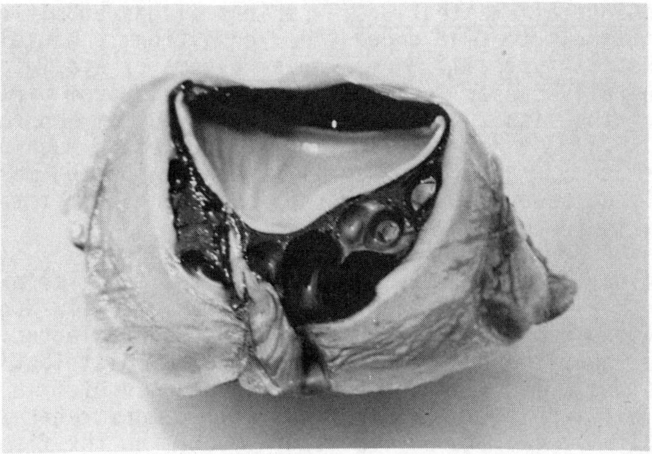

Figure 5. Accumulation of gas in choroid plexus of adult chinook salmon.

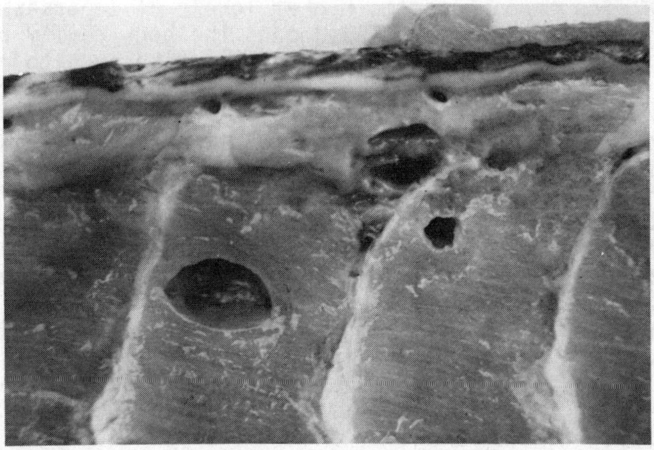

Figure 6. Muscular emphysema (cavitation) in adult chinook salmon after prolonged exposure to supersaturated water.

Pathological examinations.--Lesions of gas-bubble disease can be very difficult to find under field conditions. Acutely lethal supersaturation typically causes no emphysema or eye involvement. Death is usually caused by a massive hemostasis from air emboli, after which the fish typically sinks to the bottom and rots or is scavenged. In most cases, dead fish do not retain much evidence of what killed them. However, a small percentage of the morbid fish can usually be recovered and one should examine them internally for air emboli in the blood vascular system.

Examination of moribund or freshly dead fish for external lesions such as opercular and fin emphysema is easily done in the field provided representative specimens can be collected. It should be stressed that the occurrence, incidence and distribution of external lesions will vary widely with the species of fish affected. Also, not all fish suffering from GBD will demonstrate external lesions. It is, therefore, important to examine the fish internally for gas emboli. The afferent vessels of the gill filaments are easily examined for either single emboli or long columns of gas reaching from the base to the tip. A simple squash preparation from a moribund or recently dead fish may be examined under a dissecting scope and emboli observed in the gill filaments (Fig. 1). A mixture of blood and gas emboli may often be seen to effervesce from the ventral cut end of the branchial arch. This is particularly evident in fish exposed to the higher levels of supersaturated water. If the examiner carefully opens the body cavity of the fish without cutting major vessels, the heart may be seen to contain a mixture of blood and gas bubbles. In smaller fish, the conus arteriosis may be translucent enough to observe the emboli directly. The atrium of adult chinook salmon was often seen fully distended with gas emboli and blood (Fig. 2). Cutting the conus arteriosis of these fish frequently produces a hissing sound and effervescence of blood and gas from the cut. Although the heart and gills contain the most obvious accumulations of gas emboli, they have also been observed in other vessels. These include the coronary, meningeal, intercostal, mesenteric, testicular and ovarian vessels. Tissues such as the spleen, liver and kidney which normally contain blood sinusoids may exude a mixture of blood and gas emboli when cut. Histopathologic examination of fish from suspected supersaturation kills has often been found to be inconclusive and to offer no specific diagnosis.

Measurement of dissolved gas.--When the damage to wild fish does not produce a catastrophic mortality, the result may be extremely difficult to detect and identify. As vigor drops, a fish may fall prey to seagulls or other predators and infectious agents which tend to remove sick fish from the population. In the subacutely lethal cases, diagnosis of gas bubble disease in rivers may depend entirely upon a willingness to presume and extrapolate from saturation readings.

The easiest method of measuring dissolved gases is by the Weiss saturometer (Anon., 1972). This device is basically no more than a gas permeable tube which is sealed on one end and attached to a pressure gauge on the other end. The resulting pressure readings can be converted to various units and represent the difference (ΔP) in pressure (positive or negative) between the dissolved gas phase and the atmospheric phase. If the dissolved oxygen and temperature are known, one can estimate by calculation the levels of other gases.

Besides its ease of use, rugged and simple construction, the saturometer automatically compensates for altitude, salinity, and temperature.

A second relatively rapid and inexpensive method is that of Post (1970), which boils the gases from a known volume of water, absorbs the oxygen and carbon dioxide and estimates the nitrogen present.

There is considerable variation in the tolerance and expression of gas-bubble disease between different genera, species, families, and orders. Salmonid fishes are generally very sensitive and large mouth bass are generally quite insensitive to supersaturation. Therefore, it is impossible to predict a fish's tolerance to gas-bubble disease until it has been tested. Levels in excess of 110 percent nitrogen saturation are known to result in signs of gas-bubble disease.

ACKNOWLEDGMENTS

Contribution No. 1576 of the Rhode Island Agricultural Experiment Station. Supported in part by Sea Grant funds at the University of Rhode Island.

LITERATURE CITED

ANONYMOUS. 1973. Water Quality Criteria--1972. Prepared by: National Academy of Sciences and National Academy of Engineering. U.S. Government Printing Office, Stock No. 5501-00520, pp. 135-138.

BECKER, C. D. 1970. Temperature, timing and seaward migration of juvenile chinook salmon from the central Columbia River.

Batelle Memorial Institute, AEC Research and Development Report. BNWL-1472, UC-48, July. 21 pp.

BEININGEN, K. T. and W. J. EBEL. 1970. Effect of John Day Dam on dissolved nitrogen concentrations and salmon in the Columbia River. Trans. Amer. Fish. Soc. 99: 664-671.

BEININGEN, K. T. and W. J. EBEL. 1971. Dissolved nitrogen, dissolved oxygen, and related water temperatures in the Columbia and lower Snake Rivers, 1965-69. Data Rept. No. 56, National Marine Fisheries Service, Seattle, Washington, April 1971.

BERT, P. 1873. Influence des hautes pressions sur les poissons. Compt. Rend. Soc. Biol., Paris, 5: 160-161.

BLAHM, T. H., R. J. McCONNELL and G. R. SNYDER. 1973. Effect of gas supersaturated Columbia River water on the survival of juvenile salmonids. Final Report - Part 1. N.M.F.S. Environmental Field Station, Prescott, Oregon. 60 pp.

COUTANT, C. C. and R. G. GENOWAY. 1968. Final report on an exploratory study of interaction of increased temperature and nitrogen supersaturation on mortality of adult salmonids. A Report to U.S. Bur. of Comm. Fisheries, Seattle, Washington. Batelle Memorial Institute, Pacific Northwest Laboratories, Richland, Washington, November 28, 1968.

DANNEVIG, A. and G. DANNEVIG. 1950. Factors affecting the survival of fish larvae. Jour. Conseil. 15: 277-281.

DeMONT, J. D. and R. W. MILLER. 1971. First reported incidence of gas bubble disease in the heated effluent of a steam electric generating station. Proc. 25th Ann. Meeting, S.E. Assoc. Game and Fish Comm.

DOUDOROFF, P. 1951. Water quality requirements of fishes and effects of toxic substances, pp. 403-430. In M.E. Brown (ed.), The Physiology of Fishes. Vol. II. Academic Press, Inc., New York.

EBEL, W. J. 1969. Supersaturation of nitrogen in the Columbia River and its effect on salmon and steelhead trout. Fish. Bull. 68: 1-11.

EBEL, W. J. 1971. Dissolved nitrogen concentrations in the Columbia and Snake Rivers in 1970 and their effect on chinook salmon and steelhead trout. NOAA Tech. Report SSRF-646. 7 pp.

EBEL, W. J., E. M. DAWLEY and B. MONK. 1971. Thermal tolerance of juvenile salmon in relation to nitrogen supersaturation. Fish. Bull. 69: 833-843.

EGUSA, S. 1959. The gas disease of fish due to excess of nitrogen Jour. Fac. Fish. An. Hubs., Hiroshima U., 2: 157-182.

ELLING, C. H. and W. J. EBEL. 1973. Nitrogen supersaturation in the Columbia and Snake Rivers, a disaster or blessing in disguise. Trout. 14(2): 5 pp.

ENGELHORN, O. R. 1943. Die Gasblasenkrankheit bei Fischen. Zeits. Fischerei. 41: 297-317.

EVANS, A. and D. N. WALDER. 1969. Significance of gas micronuclei in the aetiology of decompression sickness. Nature 222: 251-252.

GERSH, I. and H. R. CATCHPOLE. 1951. Decompression sickness: Physical factors and pathological consequences, pp. 165-181. In J.F. Fulton (ed.), Decompression Sickness. Saunders, Philadelphia.

GERSH, I., G. E. HAWKINSON and E. N. RATHBUN. 1944. Tissue and vascular bubbles after decompression from high pressure atmospheres--correlation of specific gravity with morphological changes. J. Cell. Comp. Physiol. 24: 35-70.

GORHAM, F. P. 1898. Some physiological effects of reduced pressure on fish. J. Boston Soc. of Med. Sci. 3: 250-254.

GORHAM, F. P. 1901. The gas bubble disease of fish and its cause. Bull. U.S. Fish Comm., 1899, 19: 33-37.

HARRIS, M., W. E. BERG, D. M. WHITAKER and V. C. TWITTY. 1945. The relationship of exercise to bubble formation in animals decompressed to sea level from high barometric pressure. Gen. Physiol. 28: 241-251.

HARVEY, E. N. 1951. Physical factors in bubble formation, pp. 90-114. In J. F. Fulton (ed.), Decompression Sickness. Saunders, Philadelphia.

HARVEY, E. N., D. K. BARNES, W. D. McELROY, A. H. WHITELEY, D. C. PEASE and K. W. COOPER. 1944a. Bubble formation in animals. I. Physical factors. J. Cell. Comp. Physiol. 24: 1-22.

HARVEY, E. N., A. W. WHITELEY, W. D. McELROY, D. C. PEASE and D. K. BARNES. 1944b. Bubble formation in animals. II. Gas nuclei and their distribution in tissues. J. Cell. Comp. Physiol. 24: 23-34.

HARVEY, H. H. 1964. Dissolved nitrogen as a tracer of fish movements. Int. Assoc. Theoretical and Applied Limnol. 15: 947-951.

HARVEY, H. H. 1967. Supersaturation of lake water with a precaution to hatchery usage. Trans. Amer. Fish. Soc. 96: 194-201.

HARVEY, H. H. and A. C. COOPER. 1962. Origin and treatment of a supersaturated river water. International Pacific Salmon Fish Comm. Progress Rept. No. 9, 19 pp.

HARVEY, H. H. and S. B. SMITH. Supersaturation of the water supply and occurrence of gas bubble disease at Cultus Lake Trout Hatchery. Can. Fish Culturist 30: 39-46.

HEMMINGSEN, E. A. 1970. Supersaturation of gasses in water: Absence of cavitation on decompression from high pressures. Science 167: 1493-1494.

HENLY, E. 1952. The influence of the gas content of sea-water on fish and fish larvae. Rapp. Proc-Verb, Int. Conseil. Explor. Mer. 131: 24-27.

HILLS, B. A. 1967. Decompression Sickness: A study of cavitation at the liquid-liquid interface. Aerospace Med. 38: 814-817.

LANGBEIN, W. B. 1973. Water. Ann. N.Y. Acad. Sci. 216: 42-50.

MacDONALD, J. R. and R. A. HYATT. 1973. Supersaturation of nitrogen in water after passage through hydroelectric turbines at Mactaquac Dam, Saint Johns River, New Brunswick. J. Fish. Res. Bd. Canada 30: 1392-1394.

MALOUF, R., R. KECK, D. MAURER and C. EPIFANO. 1972. Occurrence of gas-bubble disease in three species of bivalve molluscs. J. Fish. Res. Bd. Canada 29: 588-589.

MARCELLO, R. A., JR. and R. K. STRAWN. 1972. The cage culture of some marine fishes in the intake and discharge canals of a steam-electric generating station, Galveston Bay, Texas. Texas A & M Univ. Sea Grant Publ. No. TAMU-SG-72-206, 267 pp.

MARSH, M. C. 1910. Notes on the dissolved content of water in its effect upon fishes. Bull. U.S. Bureau of Fisheries, 1908, 28: 891-906.

MARSH, M. C. and F. P. GORHAM. 1905. The gas disease in fishes. Report of the Bureau of Fisheries, 1904: 343-376.

MATSUE, Y., S. EGUSA and A. SAEKI. 1953. On nitrogen gas contents dissolved in flowing water of artesian wells and springs. Bull. Japanese Soc. Sci. Fish. 19: 439-444.

MERRIL, T. R., M. D. COLLINS and J. W. GREENOUGH. 1971. An estimate of mortality of chinook salmon in the Columbia River near Bonneville Dam during the summer of 1955. Fishery Bull. 68: 461-492.

MRSIC, W. 1933. Die Gasenblasenkrankeit der Fische. Ursachen Begleitarscheinumgen und Abhilfe. Zeit. Fischerei. 31: 29-67.

NAKATANI, R. E. 1969. Effects of heated discharges on anadromous fishes. Pollution. Vanderbilt Univ. Press. Nashville, Tenn.

NIMS, L. F. 1951. Environmental factors affecting decompression sickness. Part 1. A physical theory of decompression sickness, pp. 192-222. In F.J. Fulton (ed.), Decompression Sickness. Saunders, Philadelphia.

PAULEY, G. B. and R. E. NAKATANI. 1967. Histopathology of "gas bubble" disease in salmon fingerlings. J. Fish. Res. Bd. Canada 24: 867-871.

PERRIN, F. 1971. Recent advances in energy systems. Ann. N.Y. Acad. Sci. 184: 242-251.

PLEHN, M. 1924. Paktikum der Fischkrankheiten. E. Schweizerbartisch Verlagsbuchandlung, Stuttgart.

POST, G. 1970. Determination of inert gases in water. Prog. Fish Cult. 32: 170-173.

RENFRO, W. C. 1963. Gas-bubble mortality of fishes in Galveston Bay, Texas. Trans. Am. Fish. Soc. 92: 320-322.

RUCKER, R. R. 1972. Gas-bubble disease of salmonids: a critical review. Bur. Sport Fish. Wild., Tech. Paper 58, 11 pp.

RUCKER, R. R. and K. HODGEBOOM. 1953. Observations on gas bubble disease of fish. Prog. Fish. Cult. 15: 24-26.

SHELFORD, V. E. and W. C. ALLEE. 1913. The reactions of fishes to gradients of dissolved atmospheric gases. J. Exp. Zool. 14: 207-266.

SHIRAHATA, S. 1966. Experiments on nitrogen gas disease with rainbow trout fry. Bull. Fresh. Fish. Res. Lab. 15: 197-211.

SMITH, H. A., JR. 1972. N_2 threat to Pacific Northwest Fisheries? Water Spectrum 4: 41-48.

SPOTTE, S. 1973. Marine Aquarium Keeping. J. Wiley and Sons, New York.

STICKNEY, A. P. 1968. Supersaturation of atmospheric gases in the coastal waters of the Gulf of Maine. U.S. Bur. Comm. Fish. Bull. 67: 117-123.

WEBSTER, A. P. 1955. Some theoretical aspects of the use of multiple gas mixtures for deep-sea diving. In L. G. Goff (ed.), Proc. Underwater Physiol. Symposium, pp. 79-85. Publ. No. 377, Nat. Acad. Sci.-Nat. Res. Council. Washington, D.C.

WEISS, R. F. 1970. The solubility of nitrogen, oxygen and argon in water and seawater. Deep Sea Research 17: 721-735.

WEITKAMP, D. E. and M. KATZ. 1973. Resource and literature review. Dissolved gas supersaturation and gas bubble disease. Report of Seattle Marine Laboratories submitted to Northwest Utility Cooperative, Idaho Power Company, Boise, Idaho. 60 pp.

WESTGARD, R. L. 1964. Physical and biological aspects of gas-bubble disease in impounded adult chinook salmon at McNary spawning channel. Trans. Am. Fish. Soc. 93: 306-309.

WHITAKER, D. M., L. R. BLANKS, W. E. BERG, V. C. TWITTY and M. HARRIS. 1945. Muscular activity and bubble formation in animals decompressed to simulated altitudes. J. Gen. Physiol. 28: 213-233.

WOOD, J. W. 1968. Diseases of Pacific salmon, their prevention and treatment. State of Washington, Dept. Fish., Hatchery Div. 76 pp.

WORKMAN, R. D. 1963. Studies of decompression and inert gas-oxygen mixtures in the U.S. Navy. In C.J. Lambertsen and L.C. Greenbaum, Jr. (eds.), Proc. 2nd Symposium on Underwater Physiology, pp. 22-28. Publ. No. 1181, Nat. Acad. Sci.-Nat. Res. Council. Washington, D.C.

WOODBURY, L. A. 1941. A sudden mortality of fishes accompanying a supersaturation of oxygen in Lake Waubesa, Wisc. Trans. Am. Fish. Soc. 71: 112-117.

WYATT, E. J. and K. T. BEININGEN. 1971. Nitrogen gas bubble disease related to a hatchery water supply from the forebay of a high head re-regulating dam. Fish. Comm. of Oregon Res. Rept. 3: 3-12.

CHARACTERIZATION OF BENTHIC COMMUNITY RESPONSES TO ENVIRONMENTAL VARIATIONS BY MULTIPLE DISCRIMINANT ANALYSIS

Tibor T. Polgar
Martin Marietta Corporation
Environmental Technology Center
1450 South Rolling Road, Baltimore, Maryland 21227

ABSTRACT

The multiple discriminant method of community analysis is explored and developed as a technique for characterizing and comparing benthic macrofaunal assemblages. The analysis is applied to data collected in the tidal portion of the Potomac River, and the samples in discriminant space reflect the responses of community structure to substrate variations. The analysis is extended to define communities by quantifying species associations characteristic of major substrate types.

Considering substrate variations a particular example of continuous environmental change, arguments are presented for applying the method to the detection of responses to environmental perturbations. The technique requires stratified sampling designs with respect to the environmental change or gradient being investigated. Since species aggregations associated with particular conditions can be defined quantitatively, the analysis also provides for the identification of indicator species. Testing and further development of the discriminant method for the detection of environmental changes awaits the collection of appropriate field data.

INTRODUCTION

The purpose of this paper is to investigate the use of multiple discriminant analysis for characterizing benthic community organization and for detecting significant changes in ecosystem structure. Taking into account the complexity of such communities and the many factors affecting it, the most appropriate method for dealing with the allied problems of characterization and detection of change is a multivariate statistical technique. Multiple discriminant analysis was chosen because in addition to testing hypotheses, it provides for a metric canonical space (orthogonal, normalized coordinates) which is suited for the quantitative comparison of various species aggregations. It utilizes all information on the abundance and distribution of selected species and reduces the large number of variables (species) to a set of variables in much fewer dimensions (discriminant functions). The number of species to be included in such analyses is limited by the number and nature of samples available, the capabilities of computing hardware and the precision and accuracy of the numerical methods employed.

Multiple discriminant analysis will be applied to esturaine macrobenthic abundance data from various substrate habitats. The responses of community structure to continuous and discrete habitat variations are considered to be typical of other environmental variations encountered naturally and in disturbed systems. Therefore, the implication of this study is that the method is suitable for pollution impact evaluation in general, as long as the technique can be shown to be sensitive and relatively comprehensive by the inclusion of a large enough number of "important" species. From the point of view of impact assessment, the most desirable feature of the discriminant method is that communities (in a restricted sense only the species included in analysis) across habitats or other environmental gradients can be compared in bulk statistical terms with testable significance, while species-specific information is retained. The application of the discriminant method results in the partitioning of the species considered into sensitive and tolerant groups with respect to the particular condition or disturbance under investigation. Inferences of this nature are not possible from diversity measures which compound abundance and distribution characteristics ambiguously, and lose sight of the role of individual species contributions to community composition.

The benthos are well suited for use in the investigation of pollution impact because the species composition is relatively stable over time and the organisms are sessile. Estuarine benthic assemblages are composed of species with a wide spectrum of tolerances to introduced pollutants and changing conditions. Critical evaluation of changes in benthic community structure could increase understanding offthe non-lethal and long-term effects of organic input, thermal input, changes in turbidity and many other disturbances.

THEORETICAL CONSIDERATIONS AND RATIONALE

In order to evaluate the impact of pollutants or of changing environmental conditions, a method must be found for the overall characterization of the community. Without quantitative description of community state, comparison (and therefore, evaluation of change) is not possible. It is certainly impossible for any given impact study to consider the many possible biotic and biotic-physical-chemical interactions that control the function and organization of the community. These factors are recognized in conceptual terms, however, and general connections can be established among environmental factors and community organization (Carriker, 1967; Sanders, 1968).

Considering the lack of information about specific mechanisms controlling species associations and the complexity of these systems, it is clear that investigations of benthic community structure change at the present time must be correlational (statistical) and phenomenological in nature rather than causal and dynamic. Several methods of analysis are possible, but the chosen method must satisfy at least the criterion of statistical testability. The following discussion traces the traditional approaches to the problem, from which we may arrive at some techniques for improvement.

The many "baseline" pre-operational and post-operational studies one encounters in the gray literature are witnesses to a common dilemma in attempting to use resource assessment methods for impact evaluation. On one hand, there is a tendency to search for indicator species sensitive to some dominant impending environmental change. This method alone risks misinterpretation of changes occurring in the indicator species population level by attributing the observed variations solely to the factors and the species singled out for study. The complexity of links in ecosystems presents many pathways and mechanisms for change in the abundance of a given species.

A second method of assessment is to consider an assemblage or aggregation of species as a whole, reflecting the view that a community functions as a "superorganism" (Pratt, 1973). It is clear that the two points of view are complementary and not exclusive of each other. Synthesis of species and community phenomena can be accomplished theoretically by establishing the nature of dynamic processes in terms of resource partitioning. The complexion of the system as a whole will depend on fundamental species characteristics of resource utilization (niche structure), predation and competition interactions, and environmental adaptations (Schoener, 1974). In the most general sense, such syntheses would provide an explanation for species diversification in the community, and would characterize community structure.

Methodologies for achieving such a synthesis are not available for impact assessment purposes, and the efforts necessary for the accomplishment of this goal are extremely time consuming. As a substitute for rigorous methods, empirical measures of species diversity have been adopted for impact assessments. Such use of diversity measures implies that they adequately reflect community structure. The redundancies, ambiguities and arbitrary nature of diversity measures are documented (e.g., Hill, 1973; Pianka, 1971). However, the generalizations put forward by Margalef (1968), Sanders (1968) and others have provided rationales for the use of diversity measures in the detection of community changes induced by environmental variations. In relatively homogeneous environments, spatial and temporal community variations can be treated successfully by combining empirical diversity measures with general concepts of succession and diversity-stability relationships (Johnson, 1970). The same approach would also be successful in explaining community variations on a global scale.

The inadequacies and limitations of species diversity measures in characterizing communities and changes in them due to the introduction of pollutants are illustrated in the analyses of Gray (1971), Rowe (1971), Rosenberg (1973), and Lie and Evans (1973). Most demonstrated relationships between polluted and unpolluted communities have been qualitative. Further, Rowe (1971) points out that diversity measures cannot be compared among various substrate habitats. Yet habitat differences constitute a dominant and characteristic set of environmental variables determining the nature of community organization. The aquatic environment, polluted or not, is a continuum of habitats (Mills, 1969; Holland, 1974). Similarly, the distribution and abundance of species over habitats are gradational in nature. The contribution of each species to a community profile, however measured, must be determined in terms of the distribution of species within samples and their abundances, without the ambiguity displayed by diversity measures. If several communities are to be compared on the same scale, the appropriate measure would seem to be the habitat diversity (Pielou, 1972). However, most diversity measures are incapable of taking into account absences of species. As a bulk statistic for both characterization and comparison, empirical diversity measures fail in applicability.

Diversity measures, then, sometimes represent attempts to distill phenomenological information composed of the abundances of many species (variables) in order to compare such data over a spectrum of conditions (factors, treatments, stations, etc.). Whereas the concept of biological diversification is related to these measures in general, rigorous statistical characterizations and comparisons will not result from the computations. In stating the problem in terms of many variables and factors, it becomes obvious that the appropriate approach is a multivariate statistical one. To clarify this rationale, it is helpful to review the uses of various statistical techniques for hypothesis testing.

In the case of the comparison of two univariate sets of samples, we generally start with

a) two observable sets of data,

b) empirical knowledge to form hypotheses regarding the suspected differences between the two sets of samples, and

c) some knowledge of the distribution of observations and whether the samples come from the same population.

We then proceed to make estimates of distribution parameters, and apply a test statistic to determine whether differences in means exist. When the significance of differences has been determined, the estimated parameters (means) provide ready measures for the comparison of the two samples at the determined level. The test statistics might be the standardized normal, Student's t or F.

Still in the univariate case, if we desire to compare several samples representing various factors, we perform an analysis of variance: a generalization of the two group method. This is a test to determine whether there are differences in means among the groups (factors). Because the test statistics are derived from the variances within and among groups, it is important to determine the homogeneity (equality) of group variances or to maintain a symmetry in design (equal sample sizes) in order to assure the robustness of the test. The results yield knowledge of differences in means because the partition of variances gives variance ratios reflecting mean differences only (provided that group variance homogeneity is roughly satisfied). Having estimated group means and variances, and also having ascertained the significance of differences in means overall, we still have to determine which groups differ from each other and by how much. The means themselves are sufficient as measures for comparison, but generally a set of weighting factors can be applied to linear combinations of group or block means. These factors are the contrasts which allow comparison of linear combinations of means (such as differences) with determinable significance.

The next level of complexity we consider is the problem of determining differences between two sets of multivariate normal samples. Were sampling distributions of animal abundances multivariate normal, this problem could correspond to comparing two sets of macrobenthic samples, for example. A straightforward extension of the Student t-test exists, and is known as Hotelling's T^2 test.

The objective in this case is to test simultaneously hypotheses specifying the values of population means of all components of the normal vectors involved. Assuming that we can approximate multivariate normal distributions in benthic samples, for example, each observation (sample) would consist of a vector whose components are species abundance values. In these terms, we would be comparing two mean vectors whose components are the mean abundances of each species within each set of samples. The test is valid as long as the covariance (dispersion) matrices of the variables (species) in the two sets of samples differ only due to measurement errors. Thus, we are able to specify the mean vectors of the two samples and also are able to determine whether they are significantly different. To derive some unique measure of differences in the two sets of samples is a difficult matter, however. Even the components of two mean vectors present a complex set of information. We could take the vector difference and compute its magnitude as a means of putting some single value on the difference. However, the variances of each species may be different and the components of the vectors may not be independent of each other. Thus each component may have different scaling and the coordinates may not be orthogonal (statistically independent of each other). Simply put, we don't have any knowledge of the geometric properties of the sample vector space, and therefore, we are not able to define proper metrics such as Euclidean distances.

In the multivariate comparison problem for two groups, therefore, we can determine whether the groups are significantly different, but there is no simple way to put a measure on the differences. A similar situation exists if we compare several groups by multivariate analysis of variance. Although contrast vectors can be constructed and the significance of differences of combinations of group vector means can be derived, the distance measure is lacking because of the difficulties described above.

A solution to the problem of having to compare many components and to arrive at a standardized measure of overall differences in many component populations is presented by the discriminant method of analysis. The multivariate tests for two groups (Hotelling's T^2) and many groups (multivariate ANOVA) are retained structurally with a single constraint criterion in each case. When two groups are compared by discriminant analysis, the differences in mean vectors are maximized with respect to the dispersion matrices of the groups (assumed to be equal). The maximization procedure extracts a single normalized variate (canonical) onto which each multivariate sample can be mapped as a single point, containing all the information in the original components. The distances in this one-dimensional space can be computed and tested for significance. The method in the two group case is an extension of Hotelling's T^2 test.

When sampling is stratified according to a variety of factors and interactions, the second form of discriminant analysis takes on the character of multivariate analysis of variance. The constraint criterion in this case is to maximize the among-group dispersion matrix with respect to the within-group dispersion matrices. Clearly, a limitation on this procedure, by analogy, is that the dispersion matrices of various groups must be very similar. The maximization procedure now extracts several canonical variables, orthogonal to each other and normalized. The number of such variables, which are all linear combinations of the original variables (say, species), can be as many as the number of original variables or one less than the number of sample groups tested, whichever is less. Generally, however, the number of new canonical variables is much less than the number of components put into the analysis. The transformed mean vectors, whose components represent the coordinates of group centroids in discriminant space, are comparable and constitute a significant compression of all the initial information on species.

The significance of the results of the analysis is as follows:

a) the number of variables is reduced markedly (at most one in the case of two groups);

b) each sample is mapped into a space of reduced dimensions in which all coordinates are orthogonal and normalized (same scaling in all directions);

c) distances of groups from each other can be computed and tested for significance;

d) new groups encountered at some later time or collected under different conditions can be mapped into the canonical space and tested for differences; and,

e) methods can be developed from this technique to assign determined species aggregation differences to the various individual species used in the analysis.

The analysis, therefore, can classify species aggregations in a standardized manner, provide an ordination measure for them, test the differences among them, and identify those species that make the groups different. Since the experimental design is stratified according to the condition or factor being investigated, characteristic species groupings associated with significantly different factors can be defined. In addition, species which contribute greatly to group differences can be identified as "indicator" species with respect to the condition being investigated. As a viable alternative to community diversity measures, the discriminant method satisfied most criteria for resource assessment methodologies in pollution impact evaluation by providing quantitative information on both the community and species levels. Specific features of the method will be discussed following the application of the analysis to data from the Potomac River Estuary.

COLLECTION OF DATA AND FIELD METHODS

The benthic data analyzed here are derived from a field study conducted as a part of the State of Maryland's Power Plant Siting Program. The objectives of the field program included a survey of the various bottom types around the Morgantown (PEPCO) steam electric generating station in order to set up benthic sampling designs, and the testing of a hydraulically operated van Veen type grab. Equipment design, construction of the grab, and field testing were conducted under the supervision of H. Pfitzenmeyer of the Chesapeake Biological Laboratory. Sorting, identification and counting of organisms in the collections were also carried out by the CBL staff.

The hydraulic grab used in this study has several advantages over more conventional gear. The opening and closing of the grab can be initiated anytime from onboard the vessel. Because of this capability, a single grab sample may be obtained by taking short bites, so that better penetration of the bottom results. The high pressure available hydraulically also enables complete closure in most sediment types. In addition, comparable sample volumes are obtained from various substrates. The surface area of bottom sampled by the hydraulic grab was 0.124 m^2, with a mean penetration depth of 10 cm. Approximately 5 cm. depths were achieved in firm sand and shell substrates.

The specific objectives of this portion of the field program were to investigate the performance and efficiency of the hydraulic grab in various substrate types, and to determine replication requirements. Although no specific replication analyses are presented in this paper, the results of the applied statistical methods will show that 3 to 4 replicates are necessary for adequate sampling of all macrobenthic species found in three substrate types.

Figure 1 presents the geographical locations of the sampling sites. The stations are distributed over a 10 mile segment of the tidal portion of the Potomac River, in a region which could be characterized as the salt wedge boundary in the estuary at the time of sampling. Salinities decreased rapidly upstream of this river

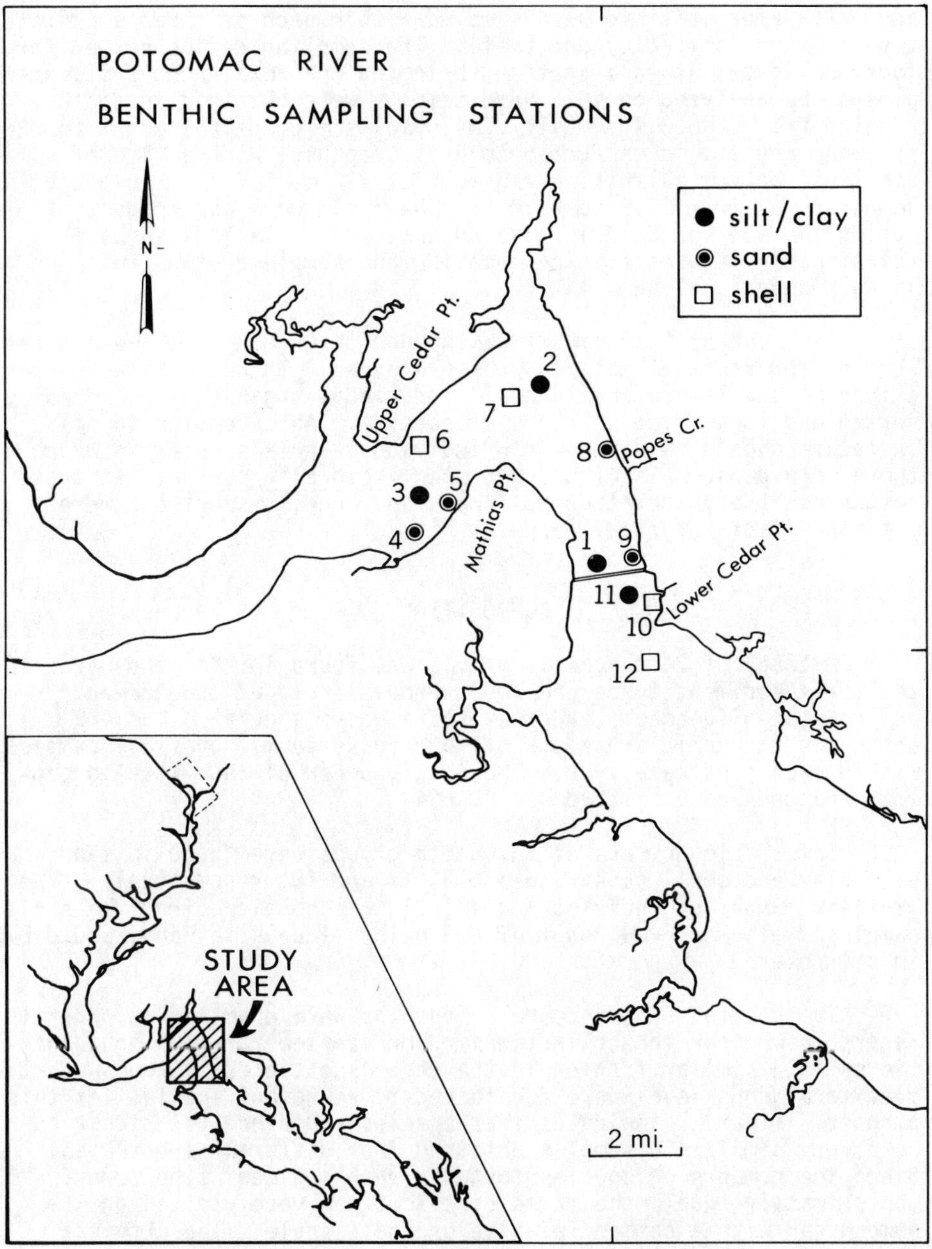

Figure 1. Location of sampling stations.

segment. Four stations were sampled within each of three substrate types: sand, silt/clay and shell. The sampling design called for four replicates at each station, bringing the total number of samples to be analyzed to 48. Sand samples were obtained in depths ranging from 6 to 8 ft.; silt/clay samples from depths of 11 to 53 ft.; and shell samples from 6 to 9 ft. depths. At the time of sampling, bottom salinities ranged from 2.7 to 5.5 ‰, and bottom temperatures from 26.9 to 29.0°C. The field work was conducted during the week of 30 July to 3 August, 1973. Table 1 shows the relationship between station location and sample designations used in subsequent analyses.

Each one of the samples was washed through a 1 mm. mesh wire screen, and residual material was preserved. The samples were examined in the laboratory, and all individual organisms were removed, sorted and identified to genus or species. An exception to this procedure should be noted for 6 individual shell samples in which there were dense mats of the bryozoan *Victorella pavida*. In these instances, the associated crustacean and annelid specimens were estimated from a partial count.

COMPOSITION

A total of 24 taxonomic groups was found in the study area in the three sediment types sampled. Groups included crustacean, polychaete, oligochaete, and several bivalve species. Table 2 shows the taxonomic breakdown of abundances summed over all samples within each substrate type. The total number of individual organisms counted and estimated was 63,664.

Comparable numbers of taxonomic groups were found in sand, silt/clay and shell substrates: 19, 17 and 18, respectively. The greatest number of individuals, 47,331, was obtained from the shell samples, followed by a count of 8,320 individuals in sand, and 8,013 in silt/clay.

The species-abundance relationships were examined in order to ascertain whether the collection method sampled rare and abundant groups in a similar fashion in the three substrates. Groups identified to a genus level were considered to represent species for this purpose. Figure 2 indicates that species-abundance relationships were very similar in samples obtained from different substrates. Since the numbers of species found in each sediment type were approximately equal, the three distributions were plotted on the same graph with a common relative ordinate scale. The abscissa represents mean densities over all samples within each substrate type, and has a logarithmic scale.

Table 1. Cross-reference table for station locations, station designations and sample (replicate) numbers.

	SAMPLE DESIGNATION	
Map Location	Sample Number*	Station Designation
4	1-4	SA1
5	5-8	SA2
8	9-12	SA3
9	13-16	SA4
1	17-20	SC1
2	21-24	SC2
3	25-28	SC3
11	29-32	SC4
6	33-36	SH1
7	37-40	SH2
10	41-44	SH3
12	45-48	SH4

*Each set of sample numbers represents a single set of replicates.

Table 2. Taxonomic breakdown of individual counts in the three sediment types.

COMPOSITION OF SAMPLE GROUPS

Index	Species	Sand	Silt/Clay	Shell	Total
1	*Cyathura polita*	166	63	44	273
2	*Edotea triloba*	4	0	0	4
3	*Chiridotea almyra*	4	1	0	5
4	*Balanus amphitrite*	0	5	6476	6481
5	*Rithropanopeus harrisi*	3	0	716	719
6	*Corophium lacustre*	153	93	22525	22770
7	*Leptochierus plumulosus*	185	104	4	293
8	*Melita nitida*	1	0	343	344
9	*Gammarus* spp.	0	8	0	8
10	*Cassadinidea lunifrons*	2	0	34	36
11	*Nereis succinea*	164	24	2818	3006
12	*Heteromastus filiformis*	434	7	5	446
13	*Eteone heteropoda*	1	0	0	1
14	*Scolecolepids viridis*	6443	5953	13261	25657
15	*Hypaniola grayi*	15	3	1	19
16	*Polydora* spp.	20	0	0	20
17	*Limnodrilus* spp.	0	1	11	12
18	*Stylochus ellipticus*	0	0	1	1
19	*Micrura* spp.	49	30	54	133
20	*Macoma balthica*	3	194	2	199
21	*Macoma phenax*	1	1	0	2
22	*Rangia cuneata*	603	1283	8	1894
23	*Brachiodontes recurvus*	0	0	44	44
24	*Congeria leucophaeta*	70	243	984	1297
	TOTALS	8320	8013	47331	63664

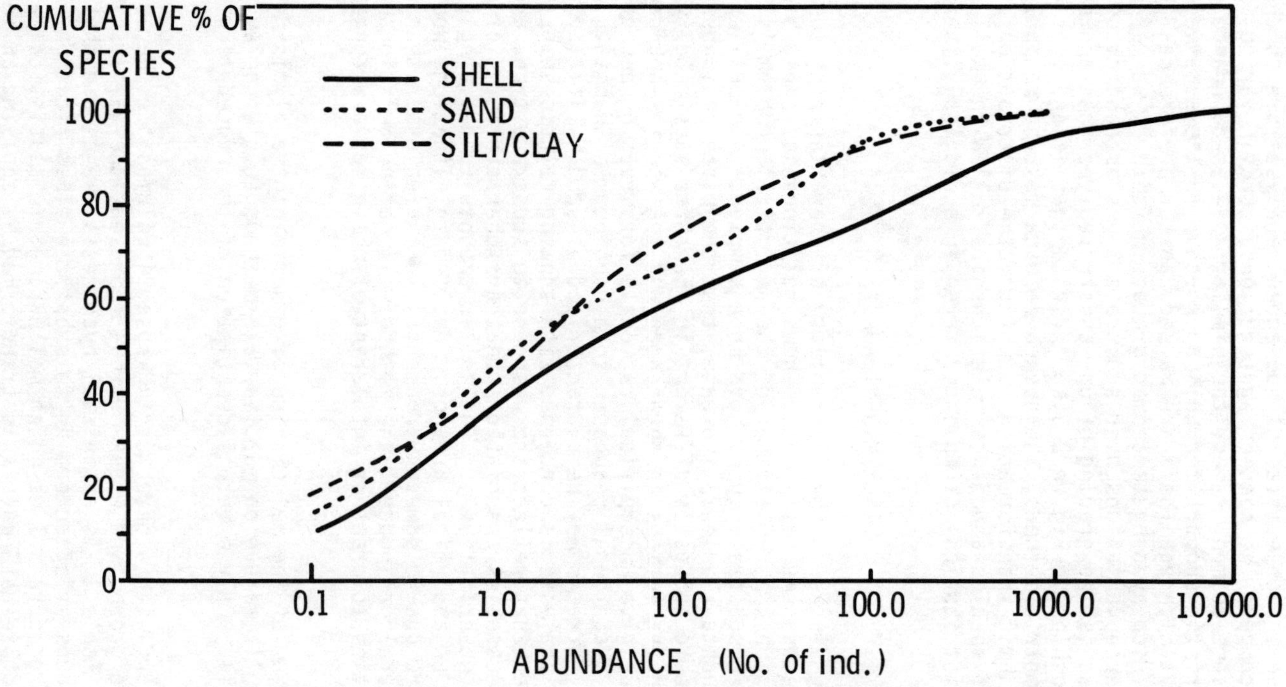

Figure 2. Species-abundance relationships in the three substrate types.

Because the objectives of subsequent analyses are to characterize and compare the species aggregations found in various substrates on some common scales, the similar species-abundance curves provide heuristic support for this approach. In the same vein, we may note that although there are marked habitat preferences and limitations displayed by taxonomic groups in Table 2, we may consider abundance variations over substrate types as continuous phenomena, as is the grading of one substrate type into another. The relationships in Figure 2 also indicate that if analyses employ normal parametric methods, the abundance data must be manipulated by a logarithmic transformation. In fact, the distribution curves suggest that species-abundance relationships follow the well-known lognormal distributions often encountered in ecological literature.

DISCRIMINANT ANALYSIS

Having discussed the desirability of characterizing species aggregations on some basis which shows relationships among biological associations, a restatement of the analytical problem is in order. Three substrate types have been sampled. A visual examination of the data (Table 2) indicates that the pattern of abundances of 24 taxonomic groups is different in the three substrate types sampled. The variations in abundances and the species-abundance relationships give an empirical rationale for treating abundance patterns as gradational responses to a continuum of habitat variations whose extremes may be characterized as sand, silt/clay and shell. The analysis is concerned with showing relationships between samples that reflect the response to substrate type, and to define those species associations that are characteristic of each environment quantitatively. The latter amounts to defining a "community" on analytical grounds. Applied to situations where responses are due to some environmental variation other than substrate, natural or induced, the method will define relationships and communities tolerant or intolerant of the given environmental condition.

A review of the mathematics of discriminant analysis may be found in the literature on multivariate methods (e.g., Cooley and Lohnes, 1971). Only a verbal description of the procedure will be given here.

We start with a data matrix whose 24 columns are the abundances of taxonomic groups and whose 48 rows represent the samples in order as shown in Table 1. As an initial hypothesis, we assume that three distinct groups exist according to substrate classification. Because the procedure relies on normal parametric statistical estimation, the entire data matrix is transformed by applying an inverse hyperbolic sine transform. This procedure centralizes the distributions and homogenizes the variances of species abundances (columns). The transformation is necessary to meet the assumptions of the method, which are essentially identical to those made in the analysis of variance.

The next step is to compute the variance-covariance (dispersion) matrix of columns (species) in the entire data set. This matrix can be regarded as the total dispersion matrix, and is analogous to total sums of squares in the univariate case. The dispersion matrices of species within each group (samples 1-16, 17-32, 33-48) are then computed, and the corresponding elements of each matrix are added to form a pooled within-group dispersion matrix. This latter matrix is analogous to the within-group of sums of squares in the univariate case. Since the total dispersion matrix can be partitioned into dispersions due to among-groups and within-groups, we may now compute the among-group dispersion matrix.

The analysis now proceeds to maximize the among-group variations with respect to the within-group variations to separate the samples in the optimal manner. This procedure is carried out by finding the latent roots (eigenvalues or characteristic roots) of the expression

$$(W^{-1} A - \lambda I) \vec{v} = 0 \tag{1}$$

where W and A are dispersion matrices within and among groups, respectively, I is an identity matrix, λ are the eigenvalues and $\vec{v}$ are the eigenvectors associated with λ. Each λ extracted (depending on the rank of $W^{-1} A$) can be used to derive an eigenvector $\vec{v}$, whose components are weights in the expression

$$y = \vec{v}' \vec{x} \tag{2}$$

where $\vec{x}$ is the vector of observations (species abundances). The quantity y is the discriminant function associated with λ. The roots λ are extracted in decreasing order of magnitude, and each one accounts for less and less variation among the groups initially hypothesized. In most cases only two discriminant functions y (in other words, a plane) are necessary to represent as many as forty groups in twenty to sixty variates (species) according to Cooley and Lohnes (1971).

The discriminant functions y are obtained in practice by multiplying the abundance of each species with the corresponding component of the eigenvector $\vec{v}$. In other words, each component of $\vec{v}$ is a weighting coefficient for the species corresponding to it in the same order. A sample can be represented as

$$\vec{X}_k = \vec{X}_k(X_{k1} \ldots\ldots\ldots\ldots X_{kp}) \tag{3}$$

and the first eigenvector derived as

$$\vec{v}_1 = \vec{v}_1\ (v_1 \ldots\ldots\ldots\ldots v_p) \tag{4}$$

when there are p species. Then, the first discriminant score for sample k is

$$y_{k1} = X_{k1}\ v_1 + X_{k2}\ v_2 \ldots\ldots + X_{kp}\ v_p \tag{5}$$

If λ_1 accounts for a significant amount of the variation among groups, say 95 percent, then all samples can be mapped as single points y_{k1} along a line, and we have succeeded in characterizing all samples in one dimension. If further discriminant functions are necessary to account for an acceptable amount of variation, then the second discriminant function is orthogonal to the first, having the same scale, and we obtain a plane for mapping each sample. The coordinates on this plane for sample k will by (y_{k1}, y_{k2}). The scales are the same in each dimension, and the two coordinates are statistically independent of each other. Since all y's are linear combinations of normal variables, they are, in fact, also standardized normal variables. Therefore, the position of one group with respect to another, that is, the distances in discriminant space, can be tested for being significantly different from zero, component by component. We have, in essence, reduced the multivariate comparison problem to several univariate problems on a component by component basis.

RESULTS AND DISCUSSION

Problems of discrimination and of hypothesis testing should be distinguished (Rao, 1952). In testing hypotheses, there is a clearly defined null hypothesis which may be accepted or rejected. Alternative hypotheses are comparatively undefined. In discrimination problems, however, a class of alternative hypotheses is presented, out of which one is chosen. The nature of the alternatives depends on the a priori probabilities associated with the initial grouping of the data. Clearly, since the method optimizes the ratio of among-group to within-group variations, the initial choice of groups forces the analysis and is critical to deciding whether there are significant differences among groups.

In the first analysis of the Potomac benthic data, the input was divided into the three substrate groups. A set of 2 discriminant functions was extracted to represent the three groups, as shown in Figure 3. As may be noted, the three groups separate well,

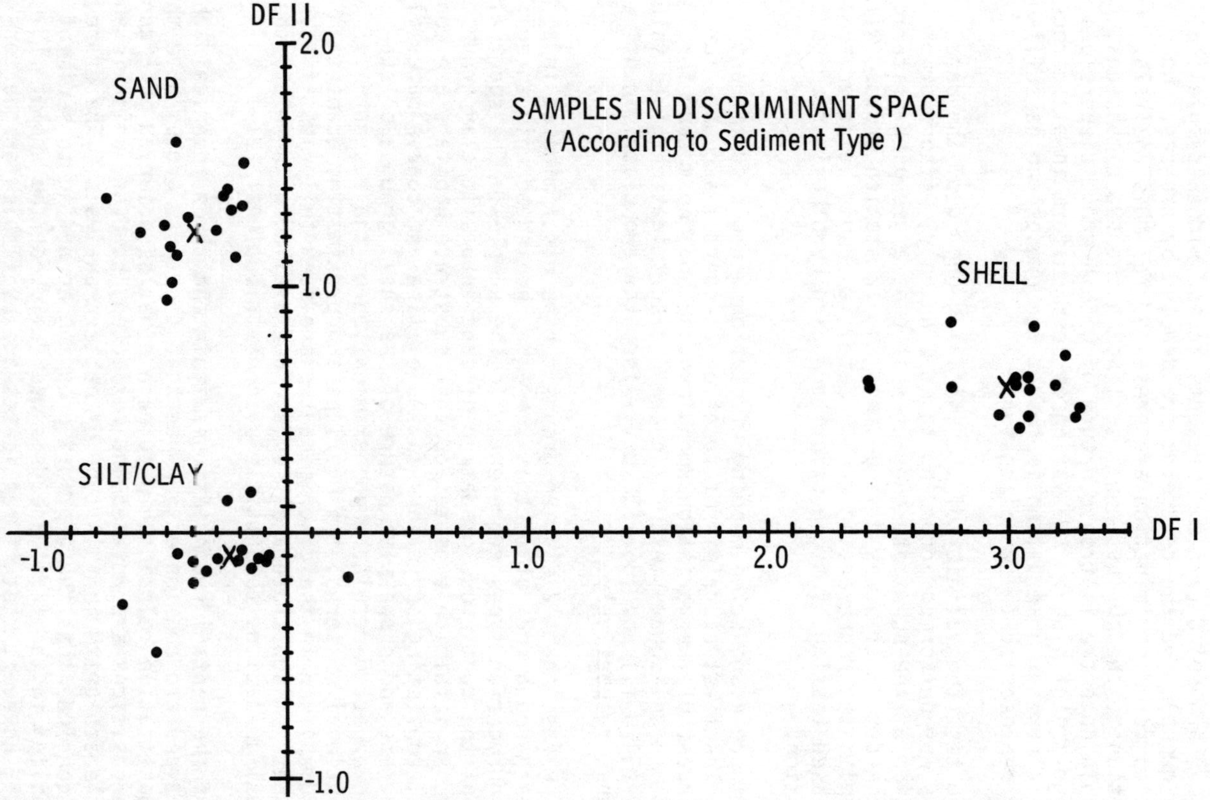

Figure 3. Discriminant analysis of the benthic data assuming an initial grouping according to sediment type (3 groups).

with sand and silt/clay groups closer to each other than to shell samples. Each solid circle represents an individual sample (16 in each group), and the centroids are denoted by crosses. No statistics for discrimination are presented for this analysis. In addition to the obvious equal scaling in the two dimensions, one should note the isotropic nature of within-group dispersions in the discriminant plane: the variances seem uniform in all directions. A further observation is that three groups can be sufficiently represented in two dimensions.

To test the validity of the initial grouping of the data, an analysis was performed according to station only, without prior assumption of the substrate characteristic as a grouping criterion. Thus, the data matrix was divided into the 12 station groups, with four replicates representing each. Here, it should be noted that an unbiased testing of the a priori grouping hypothesis required replication.

Figure 4 shows the mapping of stations into discriminant space. The first discriminant function (x-coord.) accounts for 75.5 percent of the variation among groups, and the second for 10.5 percent. A chi-square test for overall discrimination gives values of 193**(34 d.f.) and 136**(32 d.f.) for the two discriminant functions, respectively.

The fact that 12 station groups can be discriminated in a plane is a rough indication that there may be indeed only three major aggregation types represented in the samples. The spectral nature of the continuous gradation of sand into silt/clay groups displays an intuitive expectation that substrate habitats are continuous, consisting of various ratios of fine and coarse material. The discrete and dispersive nature of the shell group also supports an expected result. Upon examination of the raw data for distributional patterns, the overlaps and proximities quantitatively displayed in the discriminant plane seem fully justified. Unfortunately, substrate analyses are not available to relate the continual gradations to particle size distributions.

If the centroids of stations within the three identical substrate types are considered, we see that the three groups have the same quantitative relationship to each other as observed in Figure 3. The positions are rotated and reflected, however, which is an allowable ambiguity in this type of pattern analysis. Therefore, we may conclude that the a priori grouping according to sediment type applied in the first analysis was valid, on the whole. It is clear, however, that if replicates are available for the formation of initial groupings, the analysis will yield a finer structural definition.

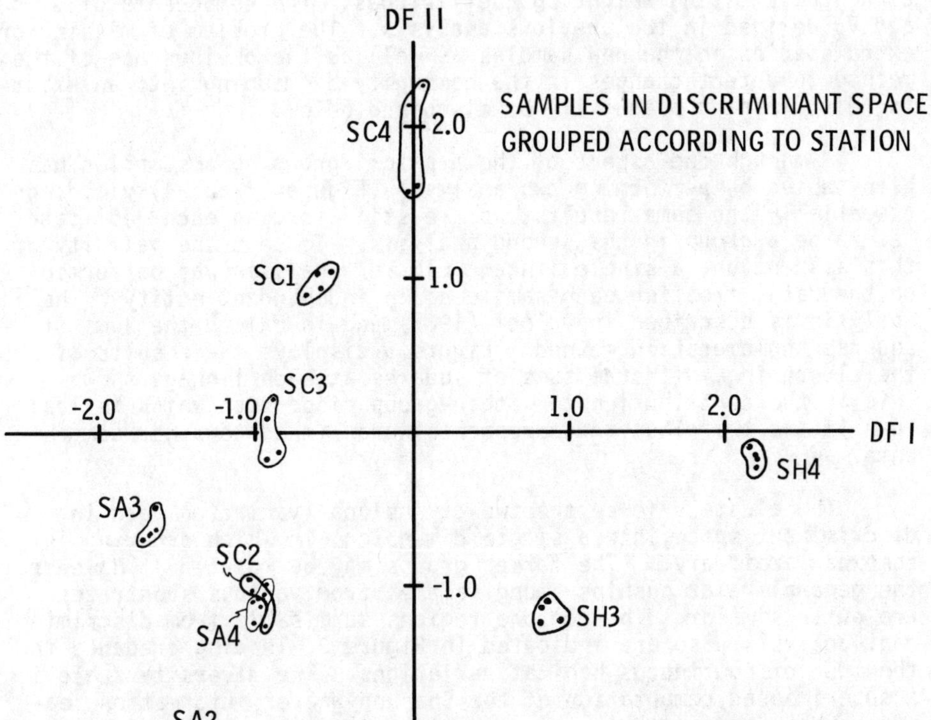

Figure 4. Discriminant analysis of the benthic data without prior assumptions based on substrate origin.

A feature of discriminant analysis not discussed before is that new samples may be mapped into the existing discriminant plane by substituting the abundances of new samples into expression (5), using the existing weighting coeffieicnes: the components of $\vec{v}_1$ and $\vec{v}_2$ derived in the previous analysis. The problem of missing or extra species in the new samples as well as the obvious use of the method to detect changes in the community by mapping into an existing discriminant plane will be discussed below.

Although one aspect of the a priori grouping assumption has been tested by performing two analyses (Figures 3 and 4) yielding essentially the same results, we are still forcing each replicate set to be a group in the second analysis. To test the validity of this assumption, a single linkage cluster analysis was performed on the data, treating each sample as an independent entity. The analysis is described in Orloci (1967) and is called the sums of squares agglomeration method. Figure 5 displays the results of the clustering, with the sums of squares at each linkage shown. This is the contribution to within-group dispersion which is less than if the two clustered components were linked to other components.

The clusters force the two-dimensional variation seen in discriminant space into a single dimension, in which ordering is somewhat arbitrary. (The forked groups may be rotated.) However, the general relationships among samples from various substrates are quite similar. The ecotone regions surmisable from discriminant analysis also are indicated in Figure 5, lending credence to the idea of continuous habitat variations. The diversity scale is a superimposed computation of the Shannon-Weaver information measure. When presented in this way, the diversity measures can be readily interpreted: high in a diverse shell substrate, less in sand, low in silt/clay, and high in ecotone mixtures. A most important result is, however, that replicate samples group remarkably well and are representative of a single station. Both this independent method and discriminant analysis underscore the need for at least three replicate samples to define quantitative community relationships in this segment of the Potomac River.

An inverse single linkage cluster analysis may be performed on the taxonomic composition in the samples. Figure 6 shows a representation of community structure with five major groups indicating abundance and preference over habitats. This analysis gives a semi-quantitative notion of distributional associations whose variations result in our capability to discriminate samples from various substrates. However, associational groups are not clearly delineated, and the analysis is heavily abundance dependent.

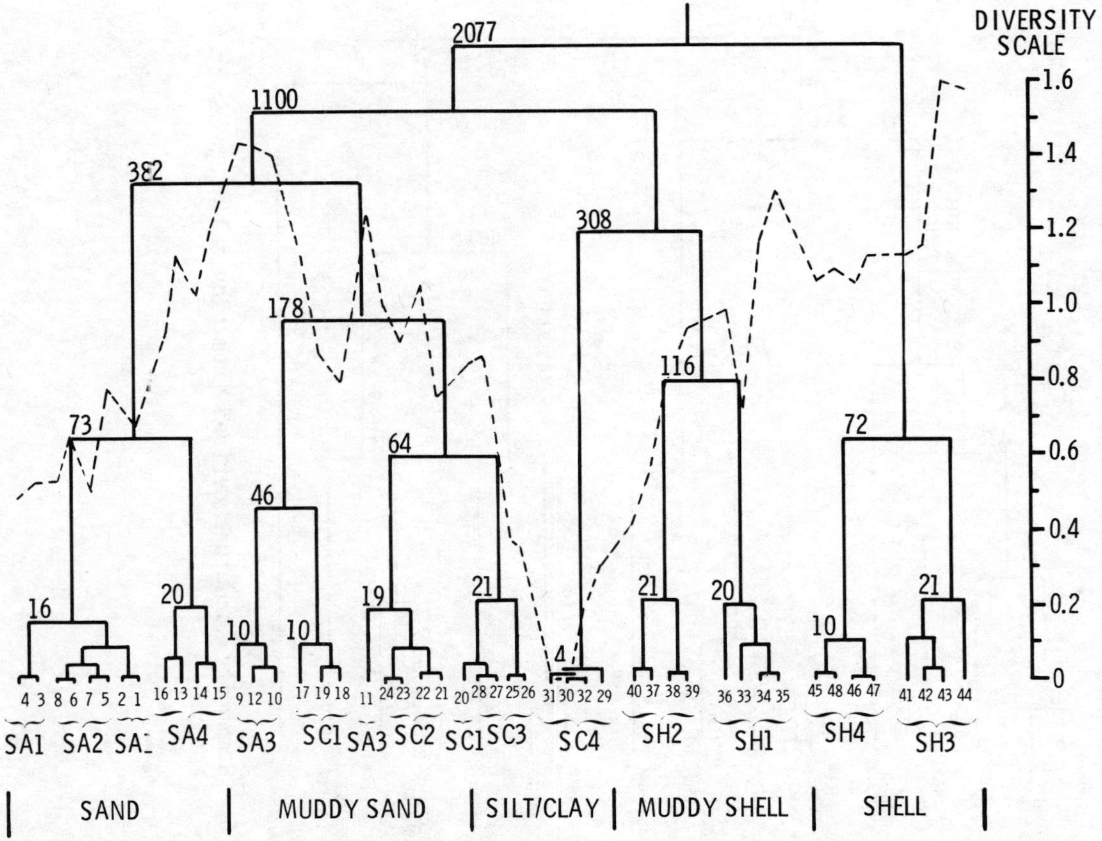

Figure 5. Single linkage clustering of samples by sums of squares agglomeration.

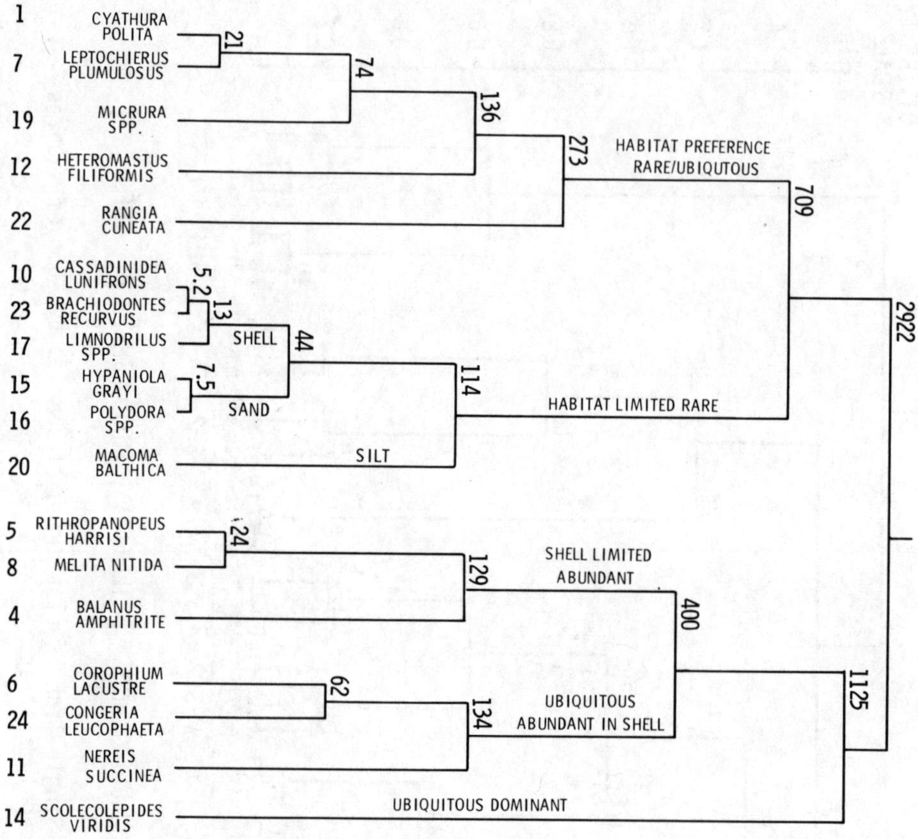

Figure 6. Community structure derived from single linkage cluster analysis.

Species 9, 18, and 21 were excluded due to their extremely low abundances.

The discriminant analysis results presented in Figure 4 can be used to define the composition of communities characteristic of each substrate type. By correlating each discriminant coordinate for all samples with the transformed abundances, we obtain a map of species aggregations which define community structure. Because the two discriminant functions account for differing amounts of variation among groups, the correlations of species abundances with discriminant scores must be weighted by the percentage of discrimination attributable to each discriminant coordinate. This procedure amounts to a canonical correlation analysis since the abundances have been transformed by a logarithmic type function and the discriminant functions are standardized normal variates.

Figure 7 shows the species in relation to their importance in contributing to the discrimination among stations shown in Figure 4. The axes are correlations between abundances and discriminant scores weighted by the amount of variation among groups attributable to each discriminant function. The groupings of species obtained from cluster analysis (Figure 6) are delineated by the superimposed, irregular boundaries. Larger absolute values of correlations result in a greater contribution of a given species to discrimination. We may conclude from this pattern of species associations that there are basically two community types in the segment of the Potomac from which the samples were obtained. Although there may be some variation in grain size, there is only a single association of organisms that grades from sand into a silt/clay community (shown on the left half plane). The shell community association is composed of the species groups in the right half plane. This map amounts to quantitative definition of the strength of species associations relatively independent of the dominance of species. The variance contributed by the distribution pattern of each species places it into either associational group. If we had stratified the samples in identical substrate types according to salinity or thermal gradients, an analogous species map would define the associations that respond strongly or weakly to the chosen environmental variable. Care must be exercised, however, in stratifying the sampling design if this is the objective because the communities may respond to several external variables at the same time.

In reference to an earlier discussion of an indicator species approach versus the community approach in impact assessment, it may be noted that in Figure 7 the outlying species (20, (1, 12), 4) are the indicators, or greatest contributors to discrimination among the three substrate conditions. The sand species group is in the

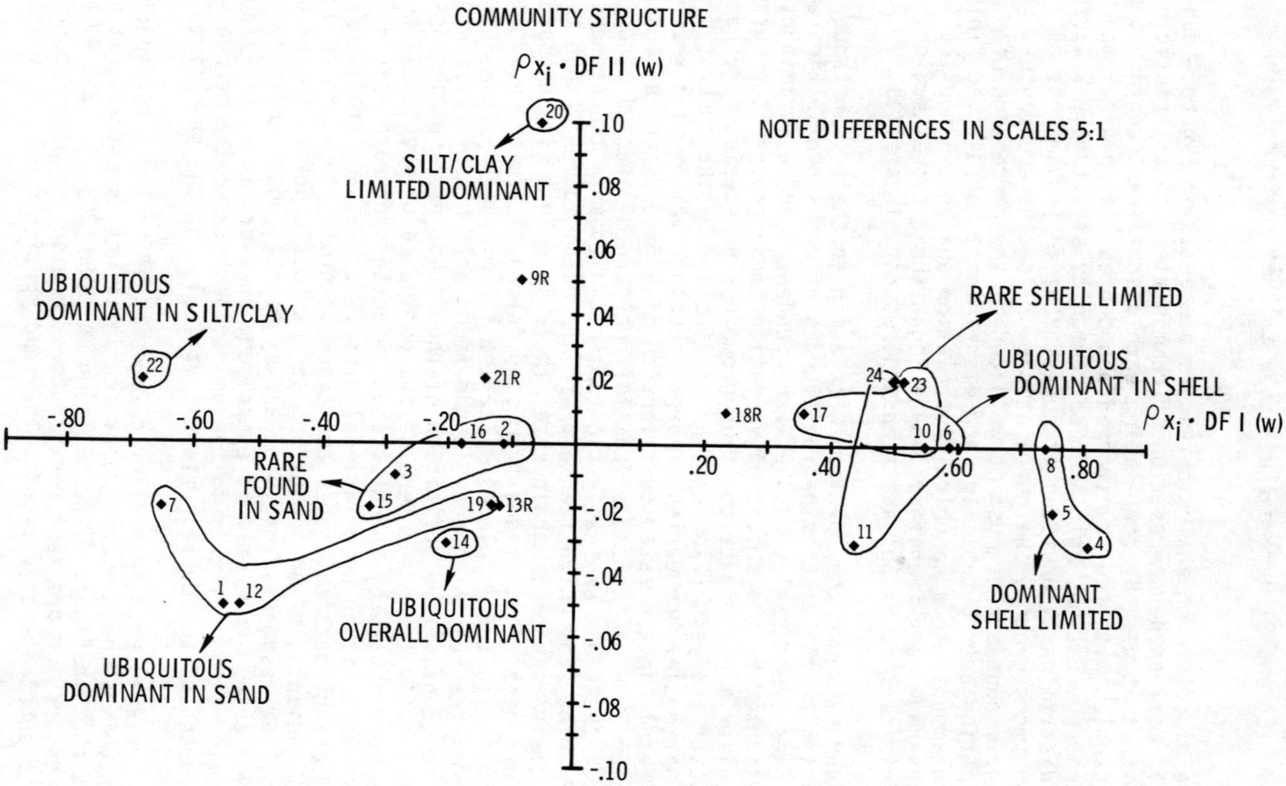

Figure 7. Community structure derived from correlations between sample discriminant scores and abundances.

lower left half plane, the silt/clay species at the upper center and the shell community in the center right half plane.

Having collected baseline data on community composition according to a sampling design stratified on the basis of an expected disturbance, post-operational samples collected at the same sites can be mapped into the existing discriminant space. Normal parametric statistics can then be employed on the discriminant coordinates, one by one, to ascertain the significance of differences. The method is yet to be tested on these kinds of data.

In anticipation of the application of the method described above, we must consider the possibility of discovering new species in samples subsequent to the collection of baseline information. In order to have a capability for including new species, we can correlate the discriminant scores of the new set of data (computed on the basis of the abundances of the species found in common with the baseline data) with the abundance of the new species. The new species can then be mapped into the existing correlation space (Figure 7), and we can determine its proximity to an existing species or species group. If such is found, the abundance of the new species may be lumped with the abundance of the most similar existing species, and the sample may be remapped still using the existing discriminant coefficients (components of $\vec{v}$). If the new species has a unique pattern with large absolute values of correlation, a new analysis of both baseline and post-operational data must be considered. Such procedures, however, have to be developed on the basis of field information.

The discriminant method of community analysis has been shown to be a powerful and sensitive tool for

a) characterizing various communities with respect to each other, and

b) defining quantitative species associations as characteristic communities.

The application of the method to the detection of community responses to conditions changing over time is yet to be tested. The range of applications can be wide, including the analysis of seasonality, the recolonization of disturbed habitats, and responses to natural or induced environmental variations.

LITERATURE CITED

CARRIKER, M. E. 1967. Ecology of estuarine benthid invertebrates: A perspective. In G.H. Lauff (ed.), Estuaries. AAAS Publication 83, Washington, D.C.

COOLEY, W. W. and P. R. LOHNES. 1971. Multivariate Data Analysis. John Wiley & Sons, New York.

GRAY, J. S. 1971. The effects of pollution on sand meiofauna communities. Thalassia Jugoslavica 7: 79-86.

HILL, M. O. 1973. Diversity and evenness: A unifying notation and its consequences. Ecology 54: 427-432.

HOLLAND, A. F. 1974. A study of the intertidal communities inhabiting sand and mud bars of the north inlet area near Georgetown, S.C., U.S.A. Ph.D. thesis, University of South Carolina.

JOHNSON, R. G. 1970. Variations in diversity within benthic marine communities. Amer. Natur. 104: 285-300.

LIE, V. and R. A. EVANS. 1973. Long-term variability in the structure of subtidal benthic communities in Puget Sound, Washington, U.S.A. Marine Biology 21: 122-126.

MARGALEF, R. 1968. Perspectives in Ecological Theory. University of Chicago Press, Chicago.

MILLS, E. L. 1969. The community concept in marine zoology, with comments on continua and instability in some marine communities: a review. J. Fish. Res. Bd. Canada 26: 1415-1428.

ORLOCI, L. 1967. An agglomerative method for classification of plant communities. J. of Ecol. 55: 193-206.

PIANKA, E. R. 1971. Species diversity. In Topics in the Study of Life: The Bio Source Book. Harper and Row, New York.

PILEOU, E. C. 1972. Niche width and niche overlap: a method for measuring them. Ecology 53: 687-692.

PRATT, S. D. 1973. Benthic fauna. In Coastal and Offshore Environmental Inventory--Cape Hatteras to Nantucket Shoals. Marine Publication Series No. 2, University of Rhode Island, Kingston.

RAO, C. R. 1952. Advanced Statistical Methods in Biometric Research. John Wiley & Sons, New York.

ROSENBERG, R. 1973. Succession in benthic macrofauna in a Swedish fjord subsequent to the closure of a sulphite pulp mill. Oikos 24: 244-258.

ROWE, G. T. 1971. The effects of pollution on the dynamics of the benthos of New York Bight. Thalassia Jugoslavica 7: 353-359.

SANDERS, H. L. 1968. Marine benthic diversity: a comparative study. Amer. Natur. 102: 243-282.

SCHOENER, T. W. 1974. Resource partitioning in ecological communities. Science 185: 27-39.

INDEX

BRIEF BIOGRAPHICAL SKETCHES OF SYMPOSIUM CONTRIBUTORS

Saul B. Saila, Professor of Oceanography and Director, Marine Experiment Station, Graduate School of Oceanography, University of Rhode Island. B.S., University of Rhode Island; M.S., Cornell University and Ph.D., Cornell University.

Kurt W. Hess, Physical Oceanographer, Marine Experiment Station, University of Rhode Island. B.S., University of California; M.S., Columbia University and Ph.D., University of Rhode Island.

Michael P. Sissenwine, Research Associate, Marine Experiment Station, University of Rhode Island. B.S., University of Massachusetts. Ph.D. requirements are being completed at the University of Rhode Island.

Jerry S. Schubel, Director, Marine Sciences Center, State University of New York at Stony Brook. B.S., Alma College; Ph.D., Johns Hopkins University.

Robert E. Hillman, Senior Marine Biologist, Batelle Columbus Laboratories, William F. Clapp Laboratories, Inc., Duxbury, Massachusetts. B.A., Hofstra College; M.A., Hofstra College and Ph.D., University of Delaware.

Robert E. Ulanowicz, Research Assistant Professor, Natural Resources Institute, Center for Environmental and Estuarine Studies, University of Maryland. B.E.S., Johns Hopkins University; Ph.D., Johns Hopkins University.

Barton C. Marcy, Technical Director for Aquatic Studies, NUS Corporation, Pittsburgh, Pennsylvania. B.S., Wake Forest University; M.S., University of Connecticut.

Thomas J. Horst, Ecologist, Stone and Webster Engineering Corp., Boston, Massachusetts. B.A., Alfred University; M.S., State University of New York at Brockport and Ph.D., Kansas State University.

J. R. Stauffer, Jr., Research Assistant, Biology Department and Center for Environmental Studies, Virginia Polytechnic Institute and State University, Blacksburg, Virginia. Ph.D. requirements being completed at Virginia Polytechnic Institute and State University.

Donald S. Cherry, Post-doctoral Appointee, Biology Department and Center for Environmental Studies, Virginia Polytechnic Institute and State University, Blacksburg, Virginia. Ph.D., Clemson University.

Kenneth L. Dickson, Associate Professor of Zoology, and Assistant Director of Center for Environmental Studies, Biology Department and Center for Environmental Studies, Virginia Polytechnic Institute and State University, Blacksburg, Virginia. Ph.D., North Texas State.

J. Cairns, University Professor and Director of Center for Environmental Studies, Biology Department and Center for Environmental Studies, Virginia Polytechnic Institute and State University, Blacksburg, Virginia. A.B., Swarthmore; M.S., Penn State and Ph.D., Penn State.

J. Richard Semple, Fish Passage Research Biologist, Department of the Environment, Fisheries and Marine Service, Halifax, Nova Scotia. B.A., St. Mary's University; M.S., University of Guelph.

Curtis L. McLeod, Fish Passage Evaluation Biologist, Department of the Environment, Fisheries and Marine Service, Halifax, Nova Scotia. M.S., University of Calgary, Ontario.

Andrew P. Beck, Research Assistant, New York University Medical Center Laboratory for Environmental Studies, Tuxedo, New York. B.S., Fairleigh Dickenson University; M.S., New York University.

Gerald V. Poje, Research Assistant, New York University Medical Center Laboratory for Environmental Studies, Tuxedo, New York. B.S. requirements being completed at New York University.

William T. Waller, Assistant Director, New York University Medical Center Laboratory for Environmental Studies, Tuxedo, New York. B.S., Kansas State University; M.S., Kansas State University and Ph.D., Virginia Polytechnic University.

Martin A. M. Hyman, Research Assistant, Marine Experiment Station, University of Rhode Island. B.A., City College of New York; M.S. requirements being completed at University of Rhode Island.

William H. Mowbray, Electronics Engineer, Marine Experiment Station, University of Rhode Island. B.S., University of Rhode Island.

Richard E. Wolke, Assistant Professor of Animal Pathology, and Director, Marine Pathology Laboratory, University of Rhode Island. B.S., Cornell University; D.V.M., Cornell University; M.S., University of Connecticut and Ph.D., University of Connecticut.

Gerald R. Bouck, Director, Western Fish Toxicology Station, USEPA, Corvallis, Oregon. B.S., Central Michigan University; M.S., Michigan State University; Ph.D., Michigan State University.

Richard Stroud, Research Associate, Department of Veterinary Medicine, Oregon State University, Corvallis, Oregon. B.S., Oregon State University; D.V.M., Washington State University and M.S. being completed at Oregon State University.

Tibor T. Polgar, Research Scientist, Martin Marietta Laboratories, Baltimore, Maryland. B.A., Amherst College; M.S., University of Rhode Island and Ph.D., University of Rhode Island.